SEPTIC TANK

OPTIONS AND ALTERNATIVES

Your Guide to Conventional, Natural and
Eco-friendly Methods and Technologies

Féidhlim Harty

Permanent Publications

Published by
Permanent Publications
Hyden House Ltd
13 Clovelly Road
Portsmouth
PO4 8DL
Tel: 01730 776 582
Email: enquiries@permaculture.co.uk
Web: www.permanentpublications.co.uk

Designed and typeset by Emma Postill

British Library Cataloguing-in-Publication Data
A catalogue record for this book is available from the British Library

ISBN 978 185623 208 1

Table of Contents

STEP A
Assessing The Current Situation

STEP B
Taking Stock: Site Characteristics, Priorities, Personal Preferences

STEP C
Making Your Selection

STEPPING THROUGH STEP C
Options Examined: Different Treatment Systems Explained

STEPPING INTO ACTION
From Decision to Implementation

About the Author

Féidhlim Harty is an environmental consultant, writer and educator living in Lahinch, Co. Clare. His company, FH Wetland Systems (established in 1996), offers guidance and consultancy on waterway repair and rewilding, eco-friendly wastewater treatment training and education, and holistic land-use management for catchment protection and regeneration.

He actively promotes environmentally regenerative solutions in the area of wastewater management and waterway protection, working with local government, state bodies, community groups and others to conserve and regenerate stream and river quality and habitat. He delivers workshops and seminars on natural treatment systems, land management methodologies and waterway enhancement measures to engineers, architects, local authorities and site assessors, as well as the general public.

Passionate about genuine environmental sustainability, he has also written *Permaculture Guide to Reed Beds*, published 2017, and *Towards Zero Waste – How to Live a Circular Life*, a practical guide to household waste minimisation, published 2019 – both by Permanent Publications. With an education in environmental science and many years of practical experience in wastewater and ecological design, his training events are a source of hope for many attendees; providing a clear direction towards a future in which we can implement practical solutions for climate action, biodiversity regeneration and community resilience.

See www.wetlandsystems.ie for details.

Acknowledgements

There are a great many people without whom this book would never have been written. Although that must be the most common opening line ever, it is always true. Above all others, it is my parents Michael and Natasha to whom I owe the most in this regard. They nurtured my fascination with the natural world, an awareness of the environmental challenges we face, and an unquenchable desire to meet those challenges to the best of my ability. My mother, along with other residents in the East Cork Harbour area and constructed wetland designer and engineer Ciaran Costello, organised the 1993 Constructed Wetland Conference in Midleton to propose an affordable, sustainable answer to the on-going pollution problems there. In many ways, small and large, my father helped me to get my fledgling business off the ground and gave me the help, practical skills and confidence to embark on those first early planting jobs. My thanks to them both, and to Ciaran and the early constructed wetland pioneers.

To all at IT Sligo who fuelled my knowledge of natural water systems, wastewater treatment processes, and what happens when you mix the two together, thank you. Over the years I think I have probably used every single skill that was on offer as part of the Environmental Science and Technology degree course.

Over the past 20 years I have had many questions from clients, engineers, architects and local authorities about wetlands, reed beds, willows and dry toilet options. Without these questions I would never have seen the need for this book. I'm also grateful to all those in the EPA and Local Authorities who have responded to my emails and queries about the *Code of Practice* and the National Septic Tank Inspection guidelines. Also to all my colleagues in the industry, for letting me bounce ideas back and forth, particularly Pamela Bartley and Ollan Herr, thank you.

Thanks also to Maddy and Tim Harland for embracing the idea of a book about septic tanks. Their dedication through the years to practical solutions for greater sustainability is truly inspiring. Also at Permanent Publications my thanks to Emma Postill for her help, suggestions and endless patience in getting the final look just right.

In the best tradition of family businesses, I roped in my daughters to help. Kate drafted the options charts that you see in each chapter, and Susie helped with the proof-reading in the final stages of the drafting process. I'm grateful to have you both in my life. Finally, I am grateful to my wife for tolerating this temporary madness and putting up with me while I spent far too much of my time writing when I should have been getting some real work done.

So, Is this Book for Me?

Having looked at the title I pick up this book. I have a septic tank problem, or think I have.

I Want to Know

(a) Whether I have a problem or not;
(b) It's ok, I can deal with it; and
(c) How to select the right choices in moving forward.

This book is a straightforward ABC to get from where you are to where you want to be.

What's in this Book?

Step A: How can I tell if I need a new system?
           Can I repair or improve my existing system?
           Is there a treatment option that can work for me?
Step B: What are my site characteristics?
           What are my budget and personal preferences?
Step C: Different options explained.
           Combining systems effectively.
           From decision to implementation.

An Unofficial Guide

This guide to sewage treatment options is not a Local Authority or EPA approved pathway towards planning permission or septic tank compliance. If you want to tick the boxes then your reference of choice should be the EPA *Code of Practice*, Wastewater Treatment and Disposal Systems Serving Single Houses (p.e.≤10) and the new EPA septic tank inspection information online.

This book refers regularly to the *Code of Practice* and septic tank inspection process, and where possible I have tried to follow the guidance in the *Code*. However the *Code of Practice* is limited to soils of good percolation and sufficient depth and to relatively conventional treatment options. Thus if you want to explore a greater selection of sustainable solutions or if you have soil that is unsuitable for percolation, then this unofficial guide may be a useful map in relatively new territory.

Bias Towards Natural and Sustainable Solutions

Let's face it, we all have biases. Mine is towards treatment systems that are natural, zero energy, recycle biomass or nutrients, and/or produce a firewood crop at the end of the year. So that's what this book gives more attention to. However, to give you a good selection of your options, all of the treatment system methodologies that I'm aware of at the time of writing are also included.

Update based on new EPA Code of Practice, 2021

The Irish EPA updated the Code of Practice on Domestic Wastewater Treatment Systems in 2021, and as such there are some changes within this book that are worth noting. The primary changes are as follows:

- The maximum percolation value (PV; formerly called t-values or p-values) permitted has increased from 90 minutes to 120 minutes (for a 25mm drop in water levels in a standardised percolation test – see pp.23, 91, 92, 96).
- The PV required for gravity distribution options for disposal to ground have been reduced from 90 minutes down to 75 minutes.
- The depth to bedrock or water table required in the code have increased. Revised figures are shown in EPA Code Table 'Minimum unsaturated soil and/or subsoil depth requirements', below:

Infiltration/treatment area	Minimum depth (m)[a]		
	GWPR R1 and R2[1]	GWPR R2[2], R2[3], R2[4] and R3[1]	GWPR R3[2]
Percolation trenches and intermittent soil filters following septic tanks	1.2	2	Not acceptable
Polishing filters following secondary systems and infiltration areas following tertiary systems (other than below)	0.9	1.2	1.8
Drip dispersal systems where the percolation value is >75. Infiltration areas following tertiary systems where the tertiary treatment system is proved to reduce E. coli to 1,000 cfu/100 ml prior to discharge to the infiltration area.[b]	0.6	0.9	1.2

[a] These depths refer to the minimum depth of unsaturated soil and/or subsoil between the point of infiltration and the bedrock and the water table. The point of infiltration is at the base of the distribution gravel in all systems, except for (a) sand filter with underlying polishing filter where it is at the base of the basal gavel layer (Figure 8.4) and (b) drip dispersal where the tubing itself is the point of infiltration.
[b] Tertiary system tested using representative secondary effluent; 90% of values complying, no value exceeding by more than 30%.

- Disposal options for primary settled or secondary treated sewage have also changed. When referring to Chapter 6, consider the new EPA disposal options sizing, in the below table, 'Infiltration/treatment area and trench length design for tertiary treatment, per PE':

Percolation values (PVs)	Pumped or underlying gravity discharge (Options 1 and 2). Area required per person (m²).	Gravity discharge into 500mm wide trenches (Option 3). Trench length required per person (m).	Low-pressure pipe distribution into 300mm wide trenches (Option 4). Trench length required per person (m).	Drip dispersal system (Option 5). Area required per person (m²).	Tertiary infiltration area (Option 6). Area required per person (m²).
3 ≤ PV ≤ 20	≥ 7.5	≥ 6	≥ 6	≥ 5	≥ 3.75
21 ≤ PV ≤ 40	≥ 15	≥ 12	≥ 12	≥ 14	≥ 7.5
41 ≤ PV ≤ 50	≥ 30	≥ 17	≥ 17	≥ 16	≥ 15
51 ≤ PV ≤ 75	≥ 50	≥ 19	≥ 19	≥ 22	≥ 25
76 ≤ PV ≤ 90	–	–	≥ 28	≥ 34	–
91 ≤ PV ≤ 120	–	–	–	≥ 54	–

Options 1, 2 and 3 are essentially as per the old EPA Code descriptions, but options 5 and 6 are new to the 2021 Code; see the code for details on these systems if your site has a PV (formerly p-value or t-value) of >90 minutes. Note that depth to bedrock and water table figures have also changed; see the EPA Code for details on these changes.

FH Wetland Systems update issued 11th September 2023. See www.wetlandsystems.ie for updates and contact information.

Introduction

After many years of being a relatively unmentionable topic, septic tanks are suddenly making the news. The dawning realisation that many of our septic tanks are actually causing water pollution seems to have been slow in coming. Now that the septic tank inspection process is being rolled out, the outcry is not so much that we should have taken better care of our drinking water supplies, of our fishing streams and our rivers and lakes, but that it might hit our pockets as we invest in repairing the damage of the past.

That said, many people have also been proactive in dealing with their wastewater; and many more are taking the septic tank inspection process as an opportunity to prioritise decisions that may have been long-fingered for want of knowing how best to tackle them. This book is written to chart a clear route forward, particularly with sustainable solutions in mind.

When I started designing and planting constructed wetland systems in the mid 1990s, septic tanks and sewage weren't exactly *de rigueur*. Although they still aren't exactly the pick-up line of the decade, they are certainly closer to the centre of conversations than they were.

Constructed wetland systems were my first love; effective wastewater treatment using no electricity, with low construction cost, eminently robust to Irish maintenance habits and providing a natural habitat to boot – what could be better? In the early stage of establishing a wastewater treatment and environmental consultancy business I viewed constructed wetlands as the one stop shop for solving all of our water pollution problems. Rivers and streams flowed fresh and clear in my daydreams, and to be fair, I have pursued the same dream even as I have divided my loyalty between other systems that have come along.

Gravel reed beds, with their more engineered look and need for additional maintenance, nonetheless offer certain advantages over constructed wetlands in certain situations. Mechanical treatment systems likewise, have certain clear advantages on some sites, despite their greater electricity usage and maintenance requirements. Most recently, zero discharge willow facilities have become an option in Ireland, offering a carbon-negative, zero discharge solution for people who want to build where there is poor percolation and no surface waters to discharge to; or those who just want the firewood and the peace of mind of having no outlet pipe anywhere.

This book should provide you with a clear overview of the different types of treatment options available and help you to choose the one most suited to your site and your needs. My bias is for natural, sustainable systems, and this is reflected in the emphasis I have placed on these systems in the book, but at the end of the day, your final choices will need to reflect your own values, priorities and site requirements, so despite my bias, this book should help to guide you towards the right system for your own particular needs.

Water Pollution in Brief

Water pollution of groundwater or surface water, such as streams and rivers, can occur whenever anything enters the water that causes a deterioration in water quality. This can include soil, silt, petrol and diesel spillages, milk, septic tank effluent, as well as rainfall runoff from farmyards, industrial yards, roads and pavements, or any number of other potential sources.

In terms of septic tank effluent, the most common causes of water pollution are direct discharges to drains and streams beside the property or unfiltered discharges into the ground via inadequate or absent percolation systems.

The main pollutants in septic tank effluent are BOD, suspended solids, ammonia, nitrates, phosphates, bacteria and other pathogens.

- BOD is short for Biochemical Oxygen Demand – and is a measure of the food value for microorganisms in the effluent. When this food is made available to bacteria, their rapid growth consumes oxygen dissolved in the water and can lead to suffocation of fish and other aquatic life, as well as a general deterioration of water quality.
- The suspended solids in effluent are literally the cloudy bits; held in suspension due to their small size. Suspended solids are often measured as an indicator of the pollution levels in effluent and are a problem in their own right in that high quantities can cause cloudiness in streams and rivers and lead to clogging of spawning beds when they settle out.
- Ammonia (NH_3) and Nitrates (NO_3) are both compounds of nitrogen. Ammonia can be toxic to fish in relatively low concentrations. The presence of ammonia can indicate recent sewage pollution since it is the first step in the nitrogen cycle after organic nitrogen in the form of proteins. Nitrates, along with phosphates, are a ready supply of nutrients for algae, which can grow in such profusion that they clog waterways and in lower quantities cause eutrophication, an over-enrichment of rivers lakes and streams, reducing habitat value and water quality.
- Phosphate (PO_4) is the phosphorus compound that is easily taken up by plants, and hence, algae. Along with nitrates they are a contributor to overall water quality deterioration in the form of eutrophication.
- Bacteria and other pathogens from sewage are many and varied. The most commonly measured bacteria in the context of sewage treatment are faecal coliforms, _E. coli_ being the most familiar. Their presence can indicate sewage or agricultural pollution. The main problem with bacteria in streams and groundwater is the potential for contamination of wells and drinking water supplies, where they can cause stomach upsets and in extreme cases, death. Other micro-organisms include some of the less pleasant fauna inhabiting the human gut; disease causing or otherwise, as well as those involved in the process of sewage treatment, that feeding frenzy that is the natural environment's response to any concentration of food. Some notable pathogens of sewage include salmonella, coliforms, parasitic worms, cryptosporidium and enteric viruses.

When septic tank effluent flows untreated into rivers, lakes or streams, it can cause contamination of drinking water supplies, nutrient enrichment and eutrophication and general overall water quality degradation and habitat damage.

Untreated discharges to groundwater also risk contaminating drinking water supplies. Also, groundwater will inevitably end up somewhere else in the course of time, as a spring or source of supply for rivers and lakes. While groundwater contamination is less easily identifiable, it can last for a long time beneath the ground and cause problems on into the future.

Why a Septic Tank Inspection Process?

The combination of a septic tank and a percolation area can be surprisingly effective at dealing with sewage IF the tank is installed in such a way as to provide adequate settlement and IF the percolation area has a sufficient surface area, an even distribution of effluent and a sufficiently deep layer of unsaturated subsoil of suitable percolation characteristics; i.e. percolation that is not too rapid and not too sluggish. If any element of the above is missing or inadequate, the whole system can fail and either clogs up visibly, causing problems for surface water, or can bypass the treatment process causing problems for groundwater.

The septic tank inspection process is an attempt to deal retrospectively with the fact that much of the soil in the country is simply not suitable for treatment by percolation and that despite guidelines and regulations to the contrary, a great many systems have not been installed correctly, even in the very recent past.

The septic tank inspection process is currently being driven by EU rulings, which endeavour to bring about a greater degree of surface water and groundwater protection within Ireland. However unpalatable the process may be, the philosophy of safeguarding our water resources is a sound one. Interestingly Cavan County Council already had a septic tank inspection process in place, and thus didn't come in for the same criticism under the EU ruling. Now that the national septic tank inspection process has started, we will hopefully see a gradual improvement in our overall environmental performance as problem sites are dealt with and a greater awareness of maintenance develops.

The opening sections in this book provide straightforward guidance on how to carry out your own inspection of your septic tank and to move if necessary towards repair or towards upgrading to a system that will work well for your site and personal preferences.

This won't replace the need for a formal inspection process, or the associated fees, but it will help you to select the system that reflects your own needs rather than just signing up to standard recommendations.

How To Use This Book

This book offers a three-step process to get from where you are to where you want to be:

   Step A is an assessment of your current system.
   Step B takes stock of your site characteristics and personal preferences.
   Step C examines the options available and looks at how to move forward.

Pick and Mix

For any treatment system you will need a number of different components. How you achieve these is up to you. Treat this book as a guide to what's on offer, and pick the systems that will work best for you. The different stages of treatment are:

- Primary (or preliminary) treatment; the solids separation stage – usually the septic tank itself, or a settlement component in a packaged treatment system.
- Secondary treatment; the oxygenation stage for reducing BOD and suspended

solids – usually provided by the percolation area, or by the air blowers in a packaged treatment system.

- Tertiary treatment; an optional extra for nutrient reduction and further 'polishing' of the effluent. This is usually an extended version of the secondary treatment stage for additional nitrogen removal, or the use of additives for phosphate removal. Sterilisation of bacteria is sometimes employed as a tertiary treatment measure where needed.
- Disposal; usually to groundwater from the percolation area, but sometimes to surface water such as a river or stream or even evapotranspiration by willow trees into the air.

There are a lot of options to choose from, many overlapping the categories above. Some systems work in complement to others, while some are mutually exclusive selections. By treating the book as a guide, you can piece together the most appropriate system for dealing with the wastewater from your home and protect your local environment.

Bonne journée.

STEP A

Assessing The Current Situation

CHAPTER 1

Initial System Check

1.1 Do I Need a New System?

This section deals with septic tanks; how they work; and how you can get yours to work better if you find that it isn't performing well. If you are building on a Greenfield site then this section won't be relevant, so you can skip ahead to the next part of the book.

In terms of answering the question of whether or not you need a new system, there are a number of initial indicators, which can provide a quick diagnosis. If you answer yes to any of the following, then you most likely need to address some shortcoming in your current system:

1. Is there effluent ponding on your lawn (or neighbour's lawn or field)?
2. Is there an overground flow directly into a field drain, stream or neighbouring property?
3. Do the sewer pipes block regularly, or is the toilet slow to flush away?
4. Do you notice obvious smells around the septic tank?
5. Does your septic tank leak?
6. Is your well, or are neighbouring wells, contaminated?

While this initial list of questions is a good beginning, answering no to each doesn't necessarily guarantee that your system works (nor indeed does answering yes mean that the situation is irreparable). Read on...

While indicators such as the age of the system, the soil type of your site and the underlying rock type all have a bearing on how your system may be performing, a thorough inspection is actually needed to fully assess it's overall health. Conversely if you are connected to a sewer and find that this option suits you well, then you can put down this book now and find something a bit more compelling to do with your time. That said there are links to eco-friendly tips at the end of the book even for people in sewered areas, so if you'd like to reduce your environmental impact, just skip to Appendix 3 now.

(See Appendix 7.1)

1.2   How Can I Tell if My Septic Tank is Working?

There are a number of standard checks that you can carry out to assess the effectiveness of your septic tank and percolation system. For many people, if the toilet flushes and the lawn isn't ponding, then the septic tank is considered fine. However it isn't always that straightforward.

Basically a septic tank should provide adequate settlement for faecal solids, toilet paper and the food bits and sludges from grey water (i.e. the water from sinks, wash hand basins, baths, showers and washing machines). After that, the percolation area should provide adequate filtration and treatment of the liquid from the septic tank through the soil, before reaching the groundwater. If either the septic tank or percolation area aren't functioning properly, then you run the risk of causing pollution, surface ponding or blocking of your system. Clearly none of these are desirable.

Before even stepping out into your garden to look at the septic tank and percolation area, two items on the list should be easy to record:

- Your toilet should flush freely. If it is slow to flush or blocks regularly, then this may indicate a problem with your septic tank or percolation area.
- If your own or neighbouring wells are known to be contaminated with sewage bacteria, then this may indicate an inadequate level of treatment, and hence a failure or inadequacy in your septic tank or percolation area performance.

Moving outdoors, the first check on the system itself is at the percolation area:

- If you don't know where your septic tank or percolation area are located, now is a good time to begin looking. If you happen to get an inspection notice, the Local Authority inspector will need to know exactly where your treatment system is located.
- The ground at and near the percolation area should be free of surface ponding.
- There should be no signs of water pollution in adjacent streams or drains. Water quality in adjacent watercourses can be analysed for chemical and microbiological contamination, but an easy visual inspection will tell a lot. For this, simply disturb the sediment in the bottom of the drain or stream with a shovel and check the colour of the mud. Dark black sediment is an indication of probable pollution by sewage or grey water. Other indicators include obvious signs of faecal material and toilet paper – but if it is that clear then you don't really need to be reading this to know you have a situation that needs attention. If there is a discharge directly into a field drain or a stream this is often indicated by green algal growth, grey 'sewage fungus' or black anaerobic mud (blacker than peaty soil, with a smell of rotten eggs or sewage).
- According to the EPA *Code of Practice*, "the percolation area should be kept free from disturbance from vehicles, heavy animals, sports activities or other activities likely to break the sod on the surface," including the use of garden tools. Note on your inspection sheet whether or not there has been surface damage, and change access or garden use if necessary.
- According to standard guidelines, percolation areas should have a distribution box to split the flow from the septic tank, as well as vent pipes rising from the end of each horizontal percolation pipe. Check that the distribution box is delivering an equal flow to each pipe, and that each vent pipe is present, intact and dry inside.

- The EPA recommendation is that the entire system be checked at least twice a year, so record the inspection date and file for safekeeping.

(See Appendix 7.2)

The next step is to examine the septic tank itself, as follows:

- Note the sludge depth and surface scum thickness. Sludge depth can be checked using the method outlined in the EPA *Code of Practice* (2009) as follows:

  1. *Use a 2m pole and wrap the bottom 1.2m with a white rag*
  2. *Lower the pole to the bottom of the tank and hold there for several minutes to allow the sludge layer to penetrate the rag, and*
  3. *Remove the pole and note the sludge line, which will be darker than the colouration caused by the liquid portion of the tank contents.*

The dark marking at the base of the pole should indicate that sludge depth is not less than 30cm from the outlet pipe T-piece. Similarly the dark brown scum layer at the tank surface should be at least 10cm above the bottom of the outlet pipe T-piece. If either of these are less than the specified distances then the tank should be desludged to prevent solids entering the percolation area. Even if the sludge and scum are not in immediate danger of being drawn directly into the percolation area, there will be a reduced retention time in the tank caused by reduced liquid volume. This will increase the volume of suspended solids entering the percolation area, leading to premature clogging. If you find that the dark mark is indistinct, then use the physical resistance of the pole to estimate the sludge depth and surface scum depth and record your findings that way.

The EPA *Code of Practice* recommends annual desludging, along with tank and percolation area inspection by the homeowner every six months. Note however that if the tank is small or the number of people using it is large, then more frequent emptying may be necessary. Conversely where only one or two people are using a large tank, then in practical terms desludging may be much less frequent than every year (see table on page 110). The EPA inspection form used by the Council inspectors checks the septic tank size and population size, so a rigid 1-year maintenance programme may not necessarily be imposed if you have a large tank to population ratio.

(See Appendix 8.3)

- Note the structural soundness and water-tightness of the tank. The best time to check for structural soundness and sewer water-tightness is when the tank is empty – however never enter a septic tank as the gasses can be extremely hazardous and lead to serious illness or death. Also avoid using electric lighting or other mains appliances to carry out checks, lest these lead to explosion of methane.

Check the empty tank for obvious signs of cracks or leaks. Listen and look for signs of groundwater seeping in through the walls or flowing in from the inlet or outlet pipes. If there is any water entering the empty tank while nobody is using the appliances in the house, it is possible that groundwater is leaking into the sewer pipe network and causing undue dilution and overall system failure or suboptimal performance. If your tank water level is below the level of the outlet pipe, this can indicate that the tank is leaking.

- Note the condition of inlet and outlet piping and of the central baffle wall. Pipes should be T-piece fittings, protruding down below the surface scum. Otherwise the scum can become disturbed and drawn through the outlet pipe where it can clog the percolation area. Check that T-piece fittings are firmly in place. A septic tank filter may be present, but is not a necessary prerequisite. If such a filter is in place, this should be removed, washed back into the tank, and replaced during each inspection. Check that the dividing wall between the two septic tank chambers is present and that the baffle is clear of debris.

In summary, if the tank needs desludging, then typically you need to engage a licensed contractor. Keep a certificate of service on file so that you can verify the date of your last desludging date. If there is relatively minor work that needs attention, carry it out as a matter of importance. If the system requires a major overhaul then read on. You may also want to contact a site assessor or environmental consultant for advice on how best to proceed. Means tested grants are also available for tanks that have been registered on time and inspected by the Local Authority and found to be in need of an upgrade.

A record sheet is included in the appendices (see Appendix 4) for recording your six-monthly septic tank system inspections. This is based on the EPA *Code of Practice* recommendations, and may be helpful during a formal septic tank inspection process, but does not replace it. The EPA inspection form used by the Local Authority tank inspectors can be found online, check Appendix 3 for the URL.

1.3   If It Looks OK, Do I Need To Do Anything At All?

If your septic tank is in place, working effectively and ticks all the boxes in Appendix 4 then it is probably working fine and providing suitable environmental protection before discharge.

Usually however, particularly if you have an older system, not all the system components are present or up to scratch. These can include T-pieces, a proper distribution box or vent pipes on the percolation area (or the percolation area itself). In such cases you may not necessarily be causing water pollution, but it is more difficult to verify compliance. In this case a detailed assessment of the site may be helpful in determining whether or not your septic tank may be causing pollution.

Sometimes you may tick all the boxes, but still be causing groundwater or surface water pollution. On sites where the initial percolation estimates were 'somewhat optimistic' and where little or no percolation actually occurs, all the component parts may be in place but a direct overflow to an adjacent drain or watercourse may also be present. In cases where you suspect poor percolation in an otherwise new site, look at any adjacent watercourses carefully to see if there are signs of sewage pollution. These are clearly identifiable by a grey scum on plants or on stones in the water or by a dark black sludge on the drain bed, noticeable when you drag a stick or spade across the bed surface. In such cases you will need to reassess your treatment options in light of these new findings. The rest of the book is designed to help you to choose a suitable system.

Where percolation areas have been installed on ground with excessive percolation, the risk of groundwater pollution is high. In this event, hope that your installer did a good job and put in the percolation area correctly to provide adequate filtration prior to reaching the water table. If you have a certificate from your engineer to sign off on the completed percolation area, or photographs of the system during construction, keep these carefully with your inspection records so that they can be used for verification if needed during the Local Authority septic tank inspection process.

If you or your neighbours have contaminated well water or suspect groundwater pollution, you may wish to examine the options available for improving the discharge quality prior to your percolation area.

CHAPTER 2

Making Good – Repairs and Improvements

2.1   If My System Doesn't Work, Can It Be Repaired or Improved?

Where a problem is encountered with any element of the septic tank self-inspection, you may be able to remedy the shortcoming yourself. Note that this does not necessarily constitute compliance with the Local Authority septic tank inspection process, but is designed to guide you towards a well functioning system that protects your local groundwater and surface water. If the list of repair work is too long, or if your site is simply unsuitable for the system you have, then you may need to choose a new system.

2.2   How Do I Go About Repairs and Improvements?

When confronted by septic tank problems, many people just don't know where to start. Some of the more common septic tank and percolation area ailments and a proposed solution to each are outlined below. The list is by no means exhaustive, but is a necessary beginning into investigating the health of your system and whether repair is a viable alternative to a full upgrade.

Background Information

Contamination of wells
Firstly consider an alternative water supply or filtration of the well water so that you have clean water while you consider your next step. Then investigate the possible sources of contamination. If your septic tank is a likely cause then investigate a new treatment system and/or percolation option. If neighbouring wells or farms are the potential source then discuss the issue with your neighbour and work towards finding a solution together. (This can sometimes be easier to put in writing than into action.)

Toilet slow to flush, or blocked
Check sewer network at manholes and tank to assess location of congestion. Rod pipes if needed. If septic tank water levels are the problem, then a deeper investigation is needed to assess the cause. For example, is the tank backing up due to a clogged percolation area or high water table etc.? A new system may be required, but only after ruling out the potential for repair.

Distribution Device

Distribution box absent between septic tank and percolation area
Common in older systems, not necessarily indicative of problems, but may signify poorly constructed or absent percolation area. No immediate remedy necessarily needed.

Leaking or providing unequal outlet flow
Repair leak. Adjust outlets to achieve equal outlet flow if possible.

Percolation Area

Percolation area location unknown
Common in older systems, not necessarily a problem, but may signify poorly constructed or absent percolation area. Location may be indicated by particularly good tree growth. Protect from vehicles and heavy use.

Surface ponding present
Indicative of poor percolation. Investigate whether storm drains allow roof water to enter sewers, and remove them if they do. Investigate a new treatment system and/or percolation option.

Vent pipes absent or damaged
If vent pipes are missing it may signify a poorly constructed or absent percolation area. If pipes are broken, repair or replace.

Vent pipes obstructed or water-logged
If pipes are clogged, free the obstruction. If they are waterlogged it is an indication of poor drainage and a new percolation option may be required.

Ground shows surface damage by vehicular activity, heavy animals, sports or other activities
Establish an exclusion area, fenced if necessary. Avoid gardening that disturbs the soil surface by using a lawn, soft fruit or no-dig techniques.

Adjacent drains and streams show signs of sewage fungus, algal growth and/or black anaerobic sludge
Establish the most likely source of pollution. If it is an upstream source then check chapter 7 on buffer zones at the end of the book. If it is your own property, then investigate repair, or replacement of your existing system.

Septic Tank

Sludge ≥30% of tank depth
Desludge the tank to avoid sludge loadings to the percolation area and consequent clogging there.

Scum layer ≤10cm from outlet pipe level
Desludge the tank to avoid solids loadings to the percolation area and consequent clogging there.

Inlet or outlet T-piece pipe absent, damaged or blocked
Repair, replace or install.

Outlet screen/filter clogged or leaking (if present)
Repair or replace if needed.

Tank not desludged in past 3 years
Check sludge depth. Empty if sludge depth is ≥30% of tank capacity. Note that particularly where grey water does not enter the tank, sludge may not accumulate and may not need annual emptying for correct performance. However if a legal requirement to desludge annually is passed, then desludging will be needed regardless of sludge depth.

Certificates absent and desludging dates unknown
Obtain and keep all new certificates. Estimate past desludging dates and note in your record sheet (see Appendix 4).

Central baffle wall absent or damaged
Repair or install a new baffle wall of concrete or heavy polypropylene or add a new tank before or after existing tank for additional settlement.

Baffle wall clogged with debris
Clear after desludging.

Water leaking in through tank walls after desludging
Repair if possible. Otherwise replace tank or install a new treatment system.

Water flowing in from percolation area after desludging
Indicative of waterlogged percolation area. Carry out more thorough investigation of the site and investigate options for a new treatment and disposal system.

Water flowing in from inlet pipe, despite no taps or appliances turned on
Indicative of leaking sewer and high water table. Sewer may need to be replaced or repaired. New treatment or percolation system may be needed based on a reassessment of the site characteristics.

Note that these pointers are an aid to getting your system working as well as you can. They are not a guarantee of compliance with the Local Authority septic tank inspection process. In general, if the works are minor, carry them out and take photographs of any process that isn't going to be visible after completion (such as re-laid pipework etc.). If the repair works are more costly, it may be well worth getting a registered site assessor to approve your plans and sign off on the finished job. Otherwise the works may need to be redone with site assessor approval and supervision in order to gain a certificate of compliance for the Local Authority inspection process. Also, where planning permission is needed, such as carrying out upgrade works rather than straightforward repairs, you may need to engage a registered site assessor to approve your proposals in order to tick the planning application boxes.

2.3   If It Can't Be Repaired or Improved, What Then?

Most issues of the septic tank itself can be repaired relatively simply. Even the installation of a new tank is not a major endeavour. The percolation area however can be a different matter altogether. For many older systems the first question that arises is the possibility that there is no percolation area at all, or that it is wholly inadequate for the task of treating and disposing of the effluent properly.

If you have soil of essentially good percolation characteristics and soil depth but have inherited a 1970s septic tank with a soak pit rather than a well laid out percolation area then you should be able to reinstall a new area without planning permission. However many councils tend to view these on a case-by-case basis so whether or not you get planning it is well worth planning your work with care and taking plenty of photos to document it as you go. Getting a site assessor, engineer or environmental consultant to check the work can be useful for a subsequent formal septic tank inspection.

If your site has poor percolation or high groundwater or rock, then straightforward percolation area replacement/improvement may not be feasible. In such cases read on and examine the various options available. This book was written with you in mind. Also, take heart, you have the company of about half the septic tank owners in the country if your system isn't working as well as it should.

2.4   Is There A System That Can Work For Me?

Yes. There is a system to suit every budget, every site, every personal preference and every legal requirement. However, these do not necessarily coincide on the same system. What you'll want to do is tick as many of these boxes as possible and move towards a system that will work best for your site, your budget, your preferences in terms of maintenance and electricity usage etc. and still remain comfortably within the limits of environmental protection and legal obligations.

STEP B

Taking Stock: Site Characteristics, Priorities, Personal Preferences

If you find that your existing system isn't doing what it should and that you need a new treatment system, percolation area or other disposal option, the next step is to carry out a careful assessment of your site. This will help to guide you towards the options that are available to you. Hand in hand with this step is to consider your own personal preferences so that from the options available you select the one most likely to be satisfactory in the long run.

CHAPTER 3

What Are My Site Characteristics?

3.1   Looking at My Site Characteristics

The two primary elements to consider are the availability of space and the capacity for final disposal of the liquid, typically via percolation to groundwater. Other details such as slope, proximity to significant features and the like, help to determine which options may be most suitable and/or what degree of treatment may be required.

The features examined during a site investigation can be divided into the following:

1. Features of the surrounding area including zoning, local housing density and prior experience in the area.
2. A visual assessment of the site, which provides information on the site size and shape, the topography and landscape, vegetation types – which in turn indicate percolation characteristics, ground condition and ease of access.
3. Minimum distance considerations such as proximity of the sewage treatment system or percolation area to dwellings, proximity to wells, roads and site boundaries. Proximity to trees, slope breaks or cuts and heritage features can also be included in this grouping.
4. After considering all of the above, the digging can start and you can begin to assess soil characteristics and percolation rate, depth to bedrock and bedrock type and the depth to water table.

The specific requirements of a site characterisation are set out in greater detail in the EPA *Code of Practice*. For planning submissions in most, if not all counties, a registered site assessor listed by the relevant council will need to carry out your site assessment. However you may wish to carry out an assessment of the site yourself to assess your treatment options or to compare the percolation value of various parts of the site prior to a formal site assessment. In this case follow the *Code of Practice*, (which is free to download from the EPA website; see the web references section) and make a careful record of your findings so that you can discuss these with your site assessor at a later date if needed.

3.2   Site Characterisation Summary

Surrounding Area

Zoning
County Development Plans or groundwater protection schemes etc. may limit the options available for sewage treatment. In sites within areas of archaeological interest, for example, the excavation for a septic tank may need the presence of an archaeologist during the excavation process.

Density of houses
If the housing density is sufficiently high it may indicate the presence of mains sewers or future plans for mains sewers. It may also prohibit additional septic tank use if the area is already at capacity. Proximity to neighbouring properties will influence the location of a treatment system and possibly the type of system chosen.

Experience of the area
Local knowledge can be very helpful in ascertaining a long-term view of the site. If every other septic tank in the area causes ponding, for example, it is likely to be indicative of heavy, unsuitable conditions for percolation.

Visual Assessment

Site size and shape
Site and shape are both important in selecting a system. Long thin sites face limitations for minimum separation distances to boundaries. For example, such sites may be more desirable for small compact mechanical secondary treatment systems than a constructed wetland of larger footprint area.

Topography and landscape
The location of a site within the landscape will indicate its suitability for some systems rather than others. Percolation in a low concave slope is likely to be less rapid than on a higher convex positions.

Vegetation types
Certain plants are indicative of wet, waterlogged conditions. Species such as alder, rush or iris are indicative of wet soils while bracken, common ragwort and creeping thistle are typically indicative of dry soil conditions.

Ground condition
The condition of the soil can vary considerably depending upon the land-use of the site. Locations within fields near gates are likely to experience ponding during heavy rains if the ground has been subjected to heavy machinery over the years. Subsoil conditions may however be quite suitable for percolation, so observation of land-use and ground condition is useful in gaining a full picture of the site.

Ease of access
All systems will require periodic access for desludging. Septic tanks and mechanical secondary treatment systems will typically require vehicle access whereas source separation systems may require only pedestrian access.

Minimum Distance Considerations

Location of adjacent dwellings or percolation areas
The closer your neighbours' dwelling, the greater the chance of minimum separation distances limiting your choice of location or choice of system within your own site.

Wells
The proximity to wells is a very important factor in the choice of a sewage treatment system. The closer the well, the greater the degree of treatment required prior to discharge. For close proximity, UV filters or other forms of sterilisation may be required to minimise bacterial contamination of the groundwater.

Proximity to surface features
Significant features include surface water, obvious ponding on the ground, exposed bedrock or steep slopes. These can place limits on the type of system and give an indication of the suitability for percolation.

Proximity to significant sites
These include such areas as Natural Heritage Areas, Special Areas of Conservation, sites of archaeological interest etc. Close proximity to such areas may limit the type or location of system chosen or specify the degree of treatment required.

Hydrological features
Streams, rivers, lakes, wells, turloughs or bogs may influence the choice of system or the degree of treatment required.

Location of dwelling on site
The location of the dwelling is important because minimum separation distances will influence where you can site your system.

Soil and Subsoil Characteristics

Soil characteristics
Soil texture, soil structure and bulk density all give an indication of the soil percolation characteristics. The presence of layering in the form of iron pans may influence drainage. Soil colouring and mottling can indicate the presence of winter water table in otherwise dry trial pits.

In sites of poor percolation, you may be able to install a constructed wetland very cost effectively, whereas a standard percolation system may be completely unsuitable.

Drainage (permeability)
Percolation test results will indicate the drainage characteristics of the site. These are important to assess the long term suitability of the soil for a percolation area. Excessive drainage leads to inadequate treatment of the discharged effluent before entry to the groundwater. Slow drainage can indicate future ponding problems over the percolation area.

Depth to bedrock and bedrock type
Depth to bedrock will influence the ease of excavation for tanks and will be a deciding factor in choosing a system and siting the percolation area (or deciding whether it will be suitable at all). The underlying bedrock can influence the water movement within

the site, and consequently the movement of effluent. Shallow soils over shale or karst limestone can lead to groundwater pollution, whereas an impermeable granite bedrock can cause extensive water-logging on the site and lead to surface water pollution in nearby streams or lakes.

Depth to water table

An adequate depth is necessary to provide treatment in a percolation area. Very high water tables may also cause problems with certain tank systems, leading to floating of tanks during desludging operations.

Prevailing wind direction

Any sewage treatment system has the potential for odours. Where too much bleach or cleaning chemicals or antibiotics enter the system, septic tank bacteria can be killed off or work less effectively. This can lead to odour generation, almost regardless of the type of system being used. The prevailing wind should ideally not carry occasional odours directly from the treatment system to your house. Some systems are more prone to odour generation than others, so wind direction is a factor in your decision making process.

3.3   Priorities and Personal Preferences

The sewage treatment system is an important aspect of the overall house/site design. It is a decision that will have long term implications for you, in terms of maintenance, on-going costs and the layout of your garden. The clearer you are about your priorities, the easier it will be to find a sewage treatment option that suits your requirements.

In addition to the purely practical aspects of site size, legal minimum separation distances and planning procurement, we all have very different attitudes to dealing with sanitation. Consequently we will all have different priorities when choosing a sewage treatment system.

Most priorities and preferences can be teased out with the following common questions:

1. How much will it cost: both to purchase/install and to run and maintain?
2. How sustainable is it: in terms of getting the sewage clean before discharge and in terms of embodied energy and electricity use?
3. How easy is it to use: will I have much work to do each year?
4. How straightforward is it to get planning: how soon can I build my house?
5. How will it fit into my garden: will it be smelly, noisy, safe, beautiful, useful?

To give you a prompt in teasing out your own preferences and priorities, the following section outlines common issues that need to be considered.

Costs

Capital cost

Consider each component of the system: supply costs, installation, completion, professional fees, commissioning, landscaping. In general terms, dry toilet systems are the most cost effective, willow facilities and systems that achieve very high tertiary treated effluent standards are the most costly, and everything else falls in between. If you can build yourself, then a treatment wetland will generally be more cost effective than an off-the-shelf system. What is the lifetime of the system – how soon will you need to outlay for the capital cost again?

Running and maintenance costs

These may include electricity costs for blowers on 24/7, pumping costs, annual desludging, component maintenance and replacement, inspection fees etc. With legal septic tank maintenance requirements coming down the tube, the rest comes down to electricity and maintenance: usually more for mechanical systems. Also check the life span of the system or system components. Filter systems, such as coconut fibre filters or gravel reed beds, generally require media replacement after a certain number of years. If you are discharging to surface waters then annual analysis and discharge licence fees are also a consideration.

Work Input

Ease of planning procurement

Some systems are simply more straightforward than others to get through planning. If you choose a system that is recommended in the *Code of Practice*, it usually makes the planning process easier and thus less expensive in terms of professional fees. This is an important consideration when selecting your preferred system, but shouldn't necessarily prevent you from selecting the system you want if you have strong reasons for selecting a system not outlined in the *Code*.

Ease of installation

This factor may be more important to you if you already have a mature garden. Off the shelf systems can usually be dropped into place and wired up to mains electricity in a single day. Willow facilities may take a week or more to install and then require follow-up work for completion the next spring. This may or may not be a deciding factor, but is worth considering in your selection process.

Ease of maintenance

Can you maintain it yourself? Can you find somebody to hand it over to? How straightforward is it to carry out? How often is maintenance or inspection required? Each system will have different answers to these questions. Mechanical systems will generally have a maintenance contract available, whereas treatment wetlands tend to be relatively self-maintaining. Source separation systems usually require on-going maintenance and care carried out by the homeowner.

Time availability

As with any project, if you have the time to do some of the work yourself, the cost can be reduced considerably. Consider both initial time input and on-going maintenance requirements. This is more applicable to septic tanks, constructed wetlands and percolation areas than to mechanical secondary treatment units, but basically the more time you can invest in the organising or installation, the less financial outlay is needed. Check the ease of operation and ease of self-maintenance of your preferred options before deciding.

Environmental Considerations

Environmental protection

How clean do you want your final effluent? This will depend to a great extent on whether you are discharging to groundwater or surface water, since surface discharges will require additional polishing prior to discharge. You may be comfortable with ticking the *Code of Practice* boxes, or you may want a much cleaner water quality again. A septic tank and percolation area is usually the minimum standard permitted (and should work fine at that if the soil conditions are right). Mechanical systems, filter systems and treatment wetlands can all achieve much the same standard, and most of these can add extra components for further pollution removal. Note that some of these can use chemicals, which will then need to be disposed of, so choose with care. At the top of the pile are systems that don't discharge anything at all, such as dry toilet systems or zero discharge willow facilities.

Embodied energy and resource consumption

Different materials have different embodied energy inputs and resource consumption. Concrete, for example, has a high embodied energy input. Plastic is an oil-based product. Gravel needs to be quarried and transported. The overall life cycle energy use of systems varies a lot, with mechanical systems bottom of the heap, treatment wetlands performing very well, and willow facilities actually net producers of energy if the wood is used for fuel.

Energy usage

If you want to keep fossil fuel and carbon footprint to a minimum, then avoid choosing a system that has a high annual electricity usage. Any pumped system will also use electricity, although less so than one that needs blowers going all year round. One exception is the willow facility, which will supply more energy yield in firewood than the pump will use.

Nutrient cycling

If you want to maximise your environmental sustainability then mixing perfectly good nutrients and organic matter with drinking water to create a potential pollution problem just doesn't make sense. Urine diversion, dry compost toilets and source separation systems are the only systems that will help us close the loop on agricultural nutrients, returning them to the land again for growing more crops. With both nitrogen and phosphorus currently sourced from finite supplies, source separation will be the way of the future, moving beyond sewage treatment to nutrient management.

Water conservation

This is both a cost issue (soon at least) and an ecological issue. If water conservation is a priority then a dry toilet system makes much more sense than pouring drinking water down the drain after compost and nutrients. Even within the category of flush toilets, dual flush and urine separation toilets can cut way down on water use.

Aesthetics

Landscape fit

The first consideration here is the legal separation distances required by the *Code of Practice*. Once that is factored in, the aesthetics of your garden layout will need to be considered. The ideal for some will be a well-planted constructed wetland, left alone for wildlife. For others it will be the minimum visible contact possible, such as a covered mechanical unit or septic tank. If you have a windy exposed site in need of shelter, perhaps a zero discharge willow facility would provide you with shelter, firewood and effluent disposal all in one. Access requirement is also a factor here; all settlement systems require access for desludging, whereas source separation systems usually only require access on foot.

Odour considerations

There are two things to consider here. Firstly, faeces smell. Evolutionarily it was probably a useful warning to keep away from potential danger of contamination. Secondly, if you add bleach to sewage and prevent appropriate bacterial breakdown, it will tend to smell worse. Septic tank and percolation systems are designed to keep odours under wraps as much as possible, and to vent gasses above the eaves of the house so they don't pose a nuisance. Packaged treatment systems likewise. Treatment wetlands are somewhat different in that they are open systems, so it is important that you treat your septic tank bacteria with loving care to help them to help you as efficiently as possible. In the context of odours, consider such factors as site size and shape; the location of your house and neighbouring houses; and the prevailing wind direction, which can influence the choice and location of the treatment system.

Assurances

Certification and guarantees

Certification and a guarantee of quality or success may be a priority for you. Most mechanical and packaged treatment systems have Agrément Certification, which certifies that in lab conditions they have achieved a certain level of treatment. Constructed systems such as percolation areas, constructed polishing filters, treatment wetlands and willow facilities do not come under the Agrément process. However these should meet certain minimum standards such as those set out in the *Code of Practice* or equivalent.

Word of mouth

If you are unsure of whether to pursue a particular system or not, it may be well worth talking with others and paying them a visit. This is particularly important for systems such as treatment wetlands, willow facilities, dry toilets or source separation systems where it's not just a box in the ground. If all you're familiar with is a septic tank, the jump to a dry toilet, for example, should only be made after lots of conversations with lots of people who use them successfully.

Other Considerations

Planning permission

Is planning permission already needed as part of your overall project, or are you dealing only with sewage treatment? To upgrade or change your sewage treatment system you need planning permission. However, if your system has failed a formal septic tank inspection, then planning is not needed for the recommended upgrade. It's unlikely that you can elect to have an inspection to bypass the planning process, however tempting it may be to speed up the process on your own site. Note that if you want a system that is not usually encountered by the local authority, such as source separation systems or willow facilities, then planning may be that bit more difficult to get.

New-build or existing house

If you are starting fresh with a new-build house project, then you can incorporate elements such as source separation systems and dual flush toilets into the bathroom design much more easily than if you are upgrading an existing system. This factor will undoubtedly influence your final choice of system.

Scales of Reference

There are many different factors to consider in the selection of a new sewage treatment set up. In order to reach a point of clarity, it may be helpful to tick a box from the following scales of reference. By clarifying our values, intentions and priorities, we can begin to see how best to move forward with selecting the right sewage treatment set-up for our site.

New-build house										Existing dwelling
System repair/ enhancement sufficient										Full system replacement required
Planning permission needed										Planning permission not needed
Impermeable clay subsoil										Free-draining shale or gravel
Site size small										Site size large
Urban area										Rural setting
Low carbon/energy a priority										Low carbon/energy not prioritised
Source separation desired										No source separation required
Water conservation a priority										Water conservation not prioritised
Maximum effluent quality a priority										Legal water quality requirement ok
Low budget a priority										Low budget not prioritised
Low maintenance preference										Low maintenance not prioritised

By ticking off the boxes in the scales of reference table above, you will have a better understanding of what system may be best for your site. Following is a summary of what systems may work for given priorities:

New-build house vs. existing dwelling
New-build opens options to water saving dual flush toilets or novel approaches such as urine diversion, which would be more costly to retrofit. Existing poses challenges to the garden, but makes it more attractive to try repair than replacement where possible.

System repair/enhancement sufficient vs. full system replacement required
Where system repair or enhancement is possible this is usually the most cost effective option. In terms of overall resource/energy consumption this may also be the most eco-friendly option.

Planning permission needed vs. planning permission not needed

Planning is generally needed to make any material changes to a sewage treatment system, including changes to location or changes to the system type. Where repair of the current system is possible, this is often the easiest approach. Where planning is needed for other site works, use this time to plan your system with care and apply for what you want rather than just the easiest thing that comes to mind.

Impermeable clay subsoil to free-draining shale or gravel

The soil percolation rate can range from very poor soakage to excessive infiltration. Where soils have a t value (percolation rate in min/25mm drop in water level) between 3 and 90 minutes, any *Code of Practice* recommendation will suffice (unless other factors come into play). For faster percolation rates, extra treatment will be needed prior to discharge. For slower percolation rates, a zero discharge willow facility, dry toilet, a surface discharge or a percolating marsh system is usually required.

Site size small to large

On small sites clearly zero discharge willow facilities and free water (open marsh) constructed wetlands will be unsuitable. On very small sites a sewer connection may be the easiest route if at all possible, or negotiating with a neighbour for land access for a percolation area. Or alternatively, a dry toilet can be a very effective space saving system. A one-acre site usually provides ample space for any treatment system you may require.

Urban or rural area

Urban sites have the advantage that sewers are usually close by. The drawback is that sites (where unsewered) tend to be smaller and in close proximity to neighbours. This may rule out most treatment wetlands as a treatment option. Proximity to wells may also be a limiting factor in siting a percolation area.

Low carbon/energy a priority or not

If you want to minimise your carbon footprint avoid mechanical systems with blowers that tend to work around the clock. After that avoid pumps altogether (willow facilities are an exception because they yield firewood which makes them net beneficial). To reduce embodied energy opt for a system that doesn't use concrete and preferably one that minimises gravel, plastic and transport as well. A dry toilet is a very low energy system in that it usually requires a lot less materials in its construction. Free water constructed wetlands built in indigenous marl clay are probably the lowest energy system that exists, with reduced budget to match.

Water conservation a priority or not

Where water conservation is specifically desired then a dry toilet is the top option, followed, after a bit of a lag, by dual flush toilets, regardless of the treatment system type. Urine separating flush toilets also minimise water use. If you don't want to change your toilet, put a plastic milk jug of water in your cistern to reduce the flow volume every time you flush, or adopt any number of water conservation measures to keep your usage to a minimum. If you want to go the extra mile then reusing grey water in your garden and polytunnel is an excellent way to conserve water. Special fittings are available for bath outlet pipes to divert clean bath or shower water to the garden, or dirty water to the sewer as needed.

Maximum effluent quality a priority or not

Essentially every toilet cistern is a spring, contributing water to your local river catchment. If you really want to keep your river healthy, then it makes sense to clean up that water as much as possible before releasing it into the groundwater. Dry toilets top the list of river-friendly options since they don't use any flush water at all and recycle all nutrients back to garden soil and plants. Even at that, a grey water system of some sort is needed to clean up sink, shower and washing machine water - which can be surprisingly polluting. A septic tank and percolation area can be ample, 'where the conditions are suitable', after that some form of secondary treatment is the next step up. Beyond that, an add-on tertiary polishing system will get it cleaner again. Any and all treatment options will offer something. What you need if this is a priority is to have several systems one after the other to achieve your desired discharge quality.

Source separation desired or not

Some people specifically want source separation of faecal matter for humanure composting or urine for a garden or agricultural fertiliser. Many don't even know that source separation is possible. If this is a priority then your main decision is whether to go for a dry toilet or a faecal or urine separation system used in conjunction with flush toilet infrastructure.

Low budget a priority or not

If your budget is approaching nil, then a home-built dry toilet system is a safe, effective and hygienic option. Most dry toilet users cite sustainability as their main reason rather than budget however. The lowest cost flush toilet option is a septic tank and percolation area. After that a constructed wetland built on indigenous marl clay is the most cost effective secondary (and tertiary where required) treatment system that exists. If budget is not limiting, then look at all your options and consider a system with flair such as one of John Todd's Living Machines (although in fairness, these are usually for high-end municipal projects with a high public profile).

Low maintenance preference or not

All treatment systems require maintenance, even if only occasional desludging. Some certainly require more maintenance than others. To minimise maintenance costs, go for a system that has no pumps, electrical components or moving parts. To minimise your own personal maintenance input you can add back in all of those things as long as you set up the relevant contracts to ensure that somebody else will take care of it when needed. Filter media units and gravel reed beds will eventually require media replacement – check with the manufacturer to see how long the media is expected to last. Dry toilets may be eco-excellent, but they require a lot more day-to-day maintenance than flush toilet equivalents. They are often less costly to maintain since you don't need to pay for desludging, but you use muscle power in it's place – either week to week or on an annual basis depending on the system you choose.

STEP C

Making Your Selection

Looking At The Options; Selecting A System; Fitting It All Together

Having investigated our existing system (if present), carried out an assessment of our site and soil characteristics and identified our priorities we are now in a position to move on to choose a system.

This chapter looks at what systems are available and at which ones are likely to be best for your site and preferences.

4.1   What Options Are There?

Because sewage treatment isn't ever really the coffee-conversation of the moment, there just isn't a high degree of knowledge about the different treatment options available. Most people have never heard of half the treatment systems that exist, and have only a nodding acquaintance with the other half. So just what options are there?

In total there must be about 100 different options available for sewage treatment in one form or another. These include treatment options for primary settlement, secondary treatment and tertiary polishing, as well as disposal of final effluent. Fortunately, these can be grouped into a number of fairly well defined categories for ease of comparison, as follows:

- New or Upgraded **Septic Tank and Percolation Systems** where the site is suitable.
- **Treatment Wetlands** for additional natural treatment where space is available.
- **Mechanical Treatment Units**, where electricity usage is deemed to be acceptable and additional treatment is required.
- **Packaged Filter Media Units** for lower electricity (or zero electricity use in some circumstances) input treatment than the mechanical units.
- **Willow Treatment Systems**, for 100% evapotranspiration, or further treatment of effluent.
- **Source Separation Systems** for diverting and recycling nutrients and biomass from the flush toilet infrastructure for increased sustainability.
- **Dry Toilet Systems** where water use is avoided and sustainability is maximised.
- **Grey Water Options** for use in tandem with dry toilet systems.
- **Disposal Options** used following most of the treatment options listed above.

Chart A: EPA *Code of Practice* options. The degree of treatment required will be a function of the site conditions

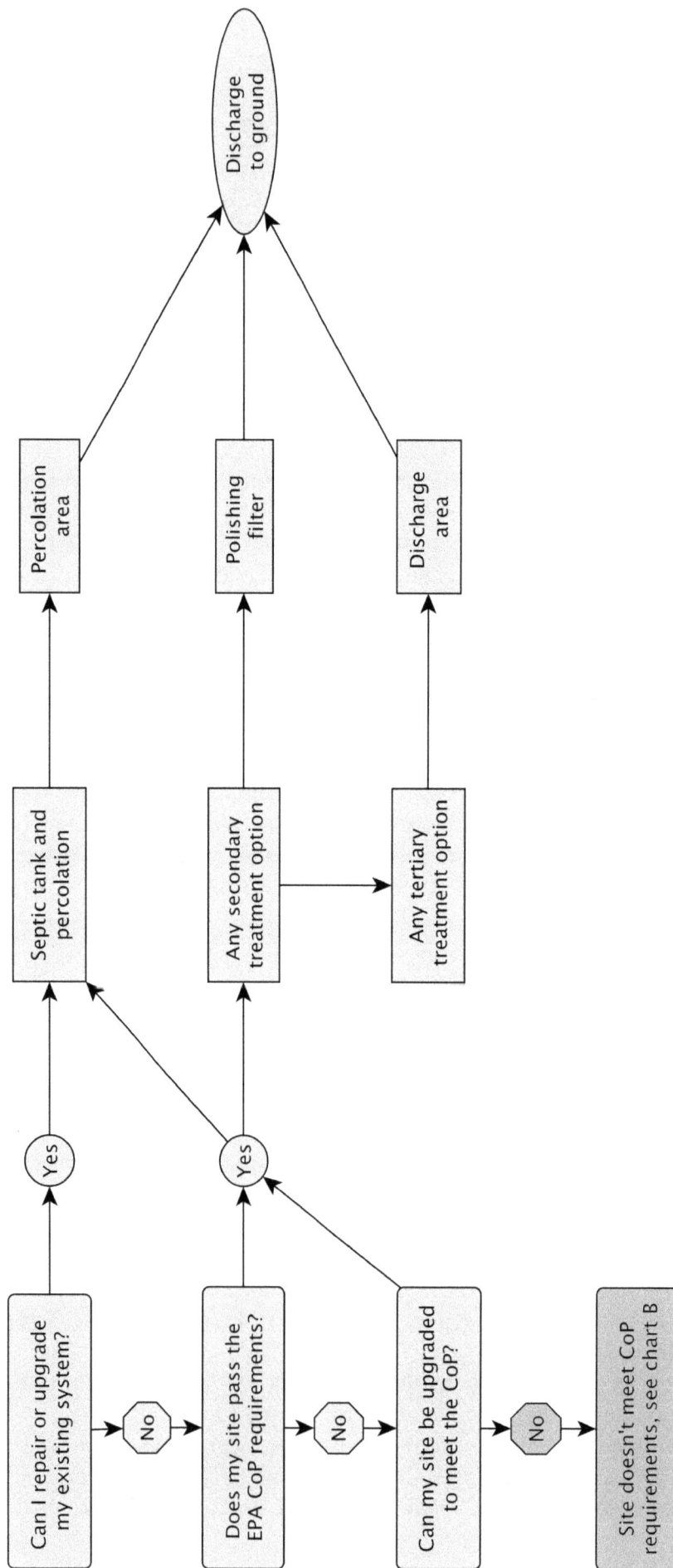

Chart B: Examples of innovative treatment methods outside the scope of the EPA *Code of Practice*

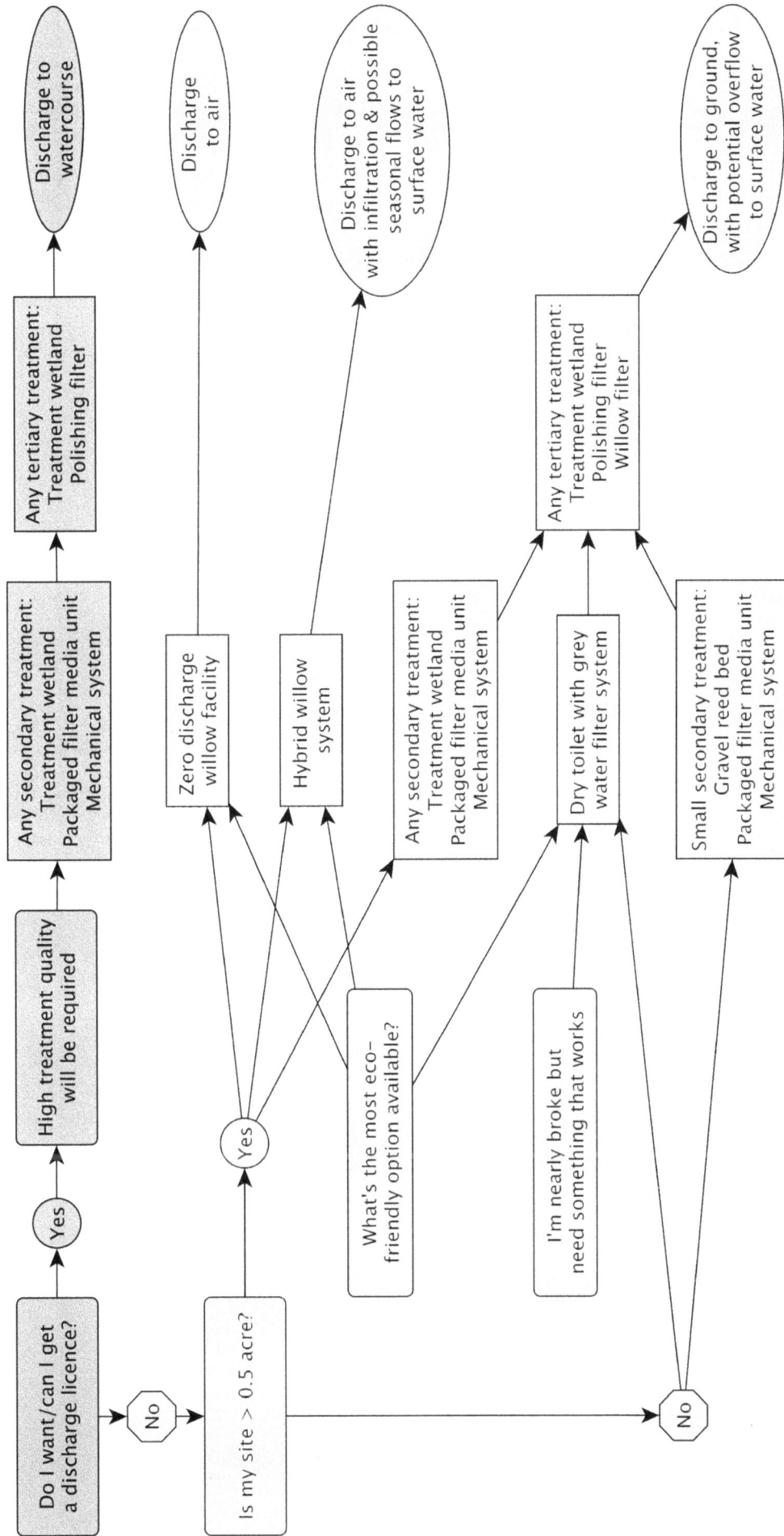

Do I want/can I get a discharge licence? → **Yes** → **High treatment quality will be required** → **Any secondary treatment:** Treatment wetland / Packaged filter media unit / Mechanical system → **Any tertiary treatment:** Treatment wetland / Polishing filter → **Discharge to watercourse**

Do I want/can I get a discharge licence? → No → **Is my site > 0.5 acre?**

Is my site > 0.5 acre? → Yes → **What's the most eco-friendly option available?**
- **Zero discharge willow facility** → Discharge to air
- **Hybrid willow system** → Discharge to air with infiltration & possible seasonal flows to surface water
- **Any secondary treatment:** Treatment wetland / Packaged filter media unit / Mechanical system → **Any tertiary treatment:** Treatment wetland / Polishing filter / Willow filter → Discharge to ground, with potential overflow to surface water

Is my site > 0.5 acre? → No → **I'm nearly broke but need something that works**
- **Dry toilet with grey water filter system**
- **Small secondary treatment:** Gravel reed bed / Packaged filter media unit / Mechanical system → **Any tertiary treatment:** Treatment wetland / Polishing filter / Willow filter → Discharge to ground, with potential overflow to surface water

4.2   Selecting the Right System for Your Site

Essentially, the easiest route to take is to follow the EPA *Code of Practice* site assessment process and in suitable site conditions, select a treatment option presented in the *Code*. A registered site assessor (whose name should be on the local authority website under the list of registered site assessors) should be able to investigate your site characteristics and guide you towards a suitable treatment option within the *Code*.

However there are number of good reasons why following the standard recommendations of the *Code* may not work. These may include the following:

- Unsuitable site conditions that fail the *Code of Practice* requirements.
- Serious budget constraints that limit the range of options available.
- Environmental preferences that lead to a desire for nutrient/biomass recycling/ composting or for maximising water conservation measures.

If any of these reasons apply to you or your site, you will need to be a lot clearer about your own options and preferences before moving forward. Bear in mind that the EPA *Code of Practice* is intended to guide the way towards an environmentally appropriate system, and while it has been essentially adopted as the final word by many councils, it specifically acknowledges that there may be other solutions available.

The flow charts on the previous pages show a simplified overview of the selection process, giving the main options available for a given site. A sewer or group sewage treatment scheme is not included in the flowchart, but should be investigated as a possibility. On Chart A, green shows standard *Code of Practice* treatment and disposal routes. On Chart B, blue shows treatment options prior to discharge to watercourses and yellow denotes options that may or may not tick the *Code of Practice* boxes but may be able to provide effective environmental protection nonetheless. Treat these charts as an example of the main treatment routes rather than a definitive.

Following is a list of some of the more common reasons why a *Code of Practice* solution may not meet your needs, with potential options beside each one. These are just a few examples, but each site is different, so work through the flow chart in tandem with the *Code of Practice* flow chart to examine your best options.

Reason	Possible Situation	Potential Options Available
Code desk study fails:	e.g. High groundwater protection requirement needed.	Dry toilet and high quality grey water treatment. Zero discharge willow facility. Extended treatment wetland area.
Code visual assessment fails:	e.g. Minimum distances determined to be inadequate.	Connect to sewer if it is a built up area and a sewer is available. Dry toilet option. Approach council with proposal for reduced distances.
On-site assessment fails:	e.g. Inadequate percolation or high groundwater table.	Treatment with tertiary polishing and surface water discharge. Zero discharge willow facility. Dry toilet option.
Serious budget constraints:	e.g. No flush toilet or inadequate, polluting system in place and no funds to improve or upgrade.	Low-cost dry toilet option with grey water infiltration area. Judicious willow planting around existing failed system.
Ecological preferences:	e.g. Desire to avoid electricity or to adopt nutrient cycling.	Gravity system such as treatment wetland, percolation system or dry toilet option. Or willow facility to produce firewood crop.

4.3   Fitting it All Together – Combining Systems Effectively

As mentioned in the introduction, we need to pick and mix the different components of the overall sewage treatment system to achieve primary settlement (or source separation), secondary treatment, tertiary polishing (where needed) and final disposal. Some options go together as a matter of course, such as septic tanks for primary settlement, followed by percolation areas for secondary treatment and subsequent disposal to groundwater. Other options take a bit more work to tease out. For example a dry toilet will provide source separation of faecal solids and urine, but will still require 'treatment' in a good compost system; and in addition, the grey water from washing machines, sinks and showers will need some sort of settlement or grease trap, treatment in a secondary treatment system such as a treatment wetland, and disposal via a soil polishing filter or planted grey water irrigation system.

Perhaps the easiest way through the selection process is not necessarily step-by-step, but by first asking what secondary system is needed, and then ascertaining what our primary treatment requirements are from that. Then in asking what disposal route we want to use, and from that stepping back to deciding whether tertiary polishing is needed or not. Essentially taking two steps forward and one step back...

Question 1 – What treatment option do I want?	Therefore what primary settlement is needed?
Percolation Area	Septic tank
Mechanical Treatment System	Typically none – usually inbuilt into the treatment system
Packaged Media Filter System	Septic tank (and pump sump where needed)
Treatment Wetland	Septic tank
Willow Facility	Septic tank and pump sump
Source Separation	Typically none (but see below for grey water)
Grey Water Treatment	Grease trap or settlement chamber

Question 2 – What disposal route do I want?	Therefore what degree of treatment is needed?
To Groundwater (where soil conditions and depths are suitable)	Percolation area or soil polishing filter.
To Groundwater (where soil depth is inadequate or percolation rate excessive)	Raised percolation area, soil or sand polishing filter and sometimes a tertiary treatment wetland or packaged tertiary treatment unit.
To Surface Water (where soil conditions are unsuitable for percolation, and surface waters are present)	Tertiary polishing via a treatment wetland, sand or soil polishing filter or packaged tertiary treatment unit.
100% Evapotranspiration (only for willow facilities)	No further treatment needed, all effluent evaporated to air.
Grey Water Recycling (typically for water conservation)	Polishing needed to keep water clean enough for reuse in toilet cisterns.
Grey Water Irrigation (typically for bath/ shower water only)	No polishing needed, but care with cosmetics, soaps and shampoos needed if reusing on food plants.

Time for Work

Given what you now know about your site characteristics and personal preferences, jot down which type of treatment option you think would be most suitable for you. With this in mind, which primary settlement system is needed to fit in with that?

Now make a note of your disposal options. Therefore what degree of treatment are you likely to need, given each disposal route you have identified?

4.4 Positioning on Site

When positioning your sewage treatment system within your site, there will be garden layout considerations and also legal minimum separation distances that you will need to observe. These legal guidelines are set out in the EPA *Code of Practice* Table 6.1 – Minimum Separation Distances In Metres, shown below:

	Septic tank, intermittent filters, packaged systems, percolation area, polishing filters (m)
Wells[1]	-
Surface water soakaway[2]	5
Watercourse/stream[3]	10
Open drain	10
Heritage features, NHA/SAC[3]	-
Lake or foreshore	50
Any dwelling house	7 septic tank 10 percolation area*
Site boundary	3
Trees[4]	3
Road	4
Slope break/cuts	4

[1] See Annex B: Groundwater Protection Response.

[2] The soakaway for surface water drainage should be located down gradient of the percolation area or polishing filter and also ensure that this distance is maintained from neighbouring storm water disposal areas or soakaways.

[3] The distances required are dependent on the importance of the feature. Therefore, advice should be sought from the local authority environment and planning sections (conservation officer and heritage officer) and/or from the Department of the Environment, Heritage and Local Government (DoEHLG), specifically the Archive Unit of the National Monuments Section and the National Parks and Wildlife Service. If considering discharging to a watercourse that drains to an NHA/SAC the relevant legislation is Article 63 of the Habitats Directive. (NHA, National Heritage Area; SAC, Special Area of Conservation.)

[4] Tree roots may lead to the generation of preferential flow paths. The canopy spread indicates potential root coverage.

* *Author note*: For soil based constructed wetland systems and most gravel reed beds, the usual FH Wetland Systems recommendation is that a minimum distance range of 30-50m be observed between the wetland and any dwelling.

Annex B of the EPA *Code of Practice* outlines the Groundwater Protection Response to be adopted for a given site. This section of the *Code* is based upon the Geological Survey of Ireland document Groundwater Protection Responses for On-Site Systems for Single Houses (DoELG/EPA/GSI, 2001).

Minimum distances between discharges to groundwater from percolation areas or polishing filters are set out based on t and p values; soil/subsoil depth above bedrock; and what classification of receiving water is present. These include Public water supply, karst features and domestic wells. The direction of flow and relative positioning of the percolation area and domestic well also has a bearing on the minimum distance requirements.

In summary, the typical recommended minimum distances are in a range between 15m and 60m. Note however that some local authorities require greater distances where a public water supply is near the site, or where streams near the site flow into such a supply. This can be up to 200m in some local authority jurisdictions.

Options Examined: Different Treatment Systems Explained

This step really looks at the background behind the different treatment options. In the next few chapters we examine each of the different treatment system types in a bit more detail and see are they really what we need. With the basic outline of your system in mind, go directly to the relevant sections in the coming chapters rather than wading through all of the information here, which by its nature will be a bit repetitive.

Chapter 5 looks at a broad range of treatment systems and sewage management approaches including conventional approaches as well as dry toilet types, grey water options and tertiary treatment systems. Chapter 6 examines the disposal options that you can employ on a site to actually get rid of the treated effluent. Chapter 7 takes a look at buffer zones, so if you can't really do much to improve a system for one reason or another – you can at least protect the environment downstream from pollution.

CHAPTER 5

Treatment System Categories

This chapter looks at the different treatment systems available in Ireland. Although it provides a broad overview of options, there are new systems emerging all the time, and hybrid systems also exist, which can complicate categorisation. In addition, bear in mind that due to the current rate of change in the area of domestic scale wastewater treatment – in terms of environmental awareness, legislation and industry developments – whole new approaches are being considered that weren't even looked at a few years ago. Not everything on the list below is necessarily included in the current *Code of Practice*, or mentioned in current legislation. Nor is everything equally environmentally sustainable. However by looking at the options you can gain an insight into the systems that are most likely to work for you.

Check on the FH Wetland Systems website for updates to this list at any time at www.wetlandsystems.ie

List of treatment options examined in this chapter:

New or upgraded Septic Tank and Percolation Systems
- Fix the old tank or percolation area
- Install a new tank or percolation area
- Pump to a polishing filter (essentially a raised percolation area)

Treatment Wetlands
- Soil-based constructed wetlands
- Integrated constructed wetlands
- Horizontal-flow gravel reed beds
- Vertical-flow gravel reed beds
- Vertical-flow sand reed beds
- Packaged tertiary treatment reed beds

Mechanical Treatment Units
- Activated Sludge (including extended aeration) systems
- Biological/Submerged Aerated Filter (BAF/SAF) systems
- Moving Bed Bioreactor (MBBR) systems
- Rotating Biological Contactor (RBC) systems
- Sequencing Batch Reactor (SBR) systems
- Membrane Bioreactor (MBR) systems.

Packaged Filter Media Units
- Coconut filter media systems
- Plastic, textile and other media systems

Willow Treatment Systems

- Zero discharge willow facilities
- Willow filter systems
- Partial evapotranspiration areas

Source Separation Systems

- Urine diversion toilets
- Urinals
- Faecal separator systems

Dry Toilet Systems

- Self-contained composting systems
- Remote composting systems
- Micro-flush systems
- Electric drying systems
- Chemical toilets

Grey Water Options

- Grey water filters and grease traps
- Planted grey water filter systems
- Grey water irrigation systems
- Grey water recycling systems

Tertiary Treatment Systems

- Soil polishing filters
- Sand polishing filters
- Treatment wetlands and reed beds
- Packaged systems

5.1   Septic Tank and Percolation Systems

A standard septic tank system comprises a settling tank and a percolation area, which in combination serve to separate solids from liquids, and to filter the liquids through the soil prior to discharge into the groundwater. In a site with ideal conditions, septic tank systems have been shown to work effectively at achieving a satisfactory level of treatment prior to entering the wider environment. Note that treated effluent quality does not necessarily mean clean water, so if you want to minimise your local environmental impact, do consider additional treatment options.

A septic tank system is composed of a number of important component parts. In any piped system, there is the sewer network itself. This must be watertight and free flowing, conveying black water from toilets and (usually) grey water from sinks, showers, washing machines etc. to the septic tank or treatment system. Access joints should be positioned at ≤30m intervals along the sewer for inspection and rodding if necessary. Also, the sewer network must convey only the grey and black water, and specifically exclude all sources of storm water from roof surfaces, yards, roads etc.

The septic tank itself has a specific set of design requirements to ensure that it works effectively at facilitating the separation of the solid fraction from the liquid fraction. Essentially, the tank must be waterproof and sufficiently robust to withstand anticipated use; must be of a sufficient size and suitable shape; must have at least two chambers for enhanced retention of sludges and scum; and must be fitted with T-pieces at the inlet and outlet to prevent scum disturbance and carry-through to the percolation area.

(See Appendix 7.2)

Following the septic tank there is a distribution box, designed to split the flow from the septic tank outlet pipe into as many different even amounts as there are pipe runs in the percolation area. Distribution boxes are often absent in older systems, and can often fail to work satisfactorily at providing an equal flow to each percolation pipe in gravity systems. Regular inspections and adjustments are recommended to remedy this.

The percolation area is a filter system that is designed to provide secondary treatment to the septic tank effluent. The percolation area is set out as a network of perforated pipes, which distribute the effluent over a gravel trench. The gravel serves as a distribution medium for the effluent, which settles on the trench base. This trench base provides the right biological environment for bacteria to form a biomat layer, which treats the effluent as it passes down through this layer into the ground, and ultimately into the groundwater below.

(See Appendix 7.1)

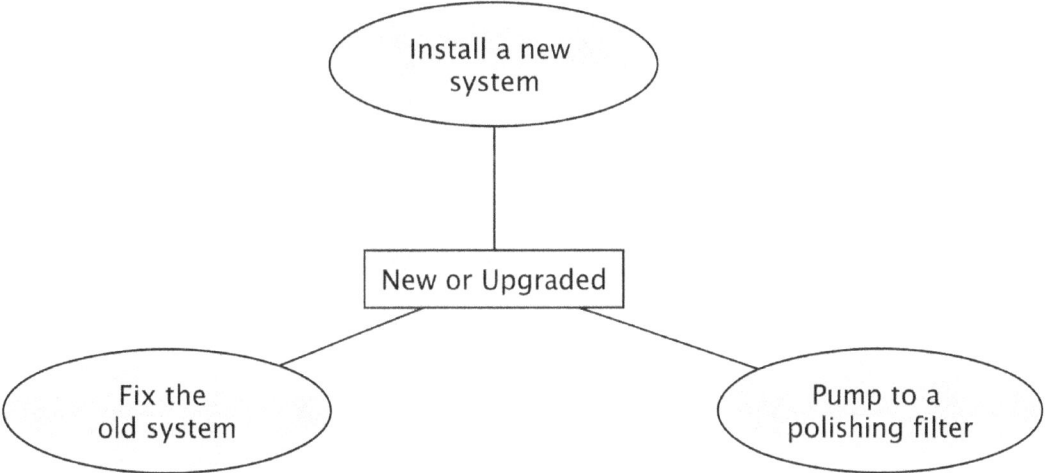

System layout within the site

Septic tank and percolation systems are typically relatively tidy in terms of layout within the site in that they are both buried beneath the ground and the percolation area is typically grassed over. Hence if your soil conditions are suitable and you are happy with having lawn where the percolation area is positioned it can be a very satisfactory treatment option from a garden layout perspective.

It is also the layout we are familiar with in Ireland, which makes it all the more acceptable. If your standard normal system is a dry toilet with a compost heap the same footprint area as a septic tank and food production as your priority, then you might find it a bit decadent to have such a large lawn, essentially doing nothing except disposing of perfectly good organic manure. However this viewpoint is still very much in the minority.

(See Appendix 8.1)

Repair/upgrade the Old Tank and/or Percolation Area

From an environmental perspective, usually if you can repair or maintain what you already have, that is often the best option. In the case of septic tanks and percolation areas, this is certainly the most cost-effective option. This can work where you have an existing tank that may be in need of upgrading rather than replacement, and where you have a percolation area in soil of suitable characteristics.

Sometimes straightforward desludging of your septic tank is all that is needed to bring your system up to standard. Sometimes it is replacement or insertion of T-pieces or baffle walls. Sometimes your percolation area just needs to be identified and formalised to provide protection from gardening or vehicle use. Other times a new percolation area may be required, in the existing soil if the site is suitable.

Pros and Cons compared to installing a new tank

Pros:
- Generally less costly than replacement.
- Repair of anything almost always has a lower embodied energy than full replacement. This is particularly the case for energy intensive materials such as concrete.
- Usually reduces or avoids the disruption of heavy machinery coming into the garden.

Cons:
- Can be difficult, unpleasant and potentially hazardous to repair septic tanks.
- Does not necessarily guarantee full compliance with the EPA *Code of Practice*, although it may do so.
- Tank may be discovered to be beyond repair after spending time and money on trying.

Install a New Tank and/or Percolation Area

Where the old septic tank or percolation area are inadequate or non-existent, but site characteristics permit their use, a new system may provide the easiest route to an effective on-site treatment solution.

Pros and Cons compared to other treatment options

Pros:
- Septic tank and percolation systems are relatively inexpensive and function without electricity where site gradients permit.
- Where suitable soil conditions and depths exist they can provide effective treatment from an environmental perspective, providing roughly the equivalent of secondary treatment.

Cons:
- Not all soils are suitable for septic tank and percolation systems.

- Proximity to wells, vulnerable aquifers or watercourses can render this form of treatment inadequate.
- Personal preferences may dictate that secondary or tertiary treatment standard first is more desirable than treatment provided by percolation alone.

Pump to a Polishing Filter

Where the existing percolation area is ponding or where the depth to bedrock or groundwater is inadequate, it is possible that construction of a raised percolation mound may be a suitable route forward. This will only work where the underlying subsoil has sufficient percolation characteristics to allow water to disperse into the soil after treatment in the raised polishing filter. A new pump sump and pumped feed to the new raised mound are both needed in this instance.

Pros and Cons compared to other treatment options

Pros:
- This option can take advantage of the plus sides of soil percolation on soils with high groundwater table or bedrock, while ticking legislative boxes and providing environmental protection.
- A pumped feed is less energy intensive than a secondary treatment system that requires constant energy inputs for 24 hours a day. Pumps at least only consume power when there is water present to deliver to the raised percolation mound.

Cons:
- A pump is typically needed to get water up to the raised percolation mound, so electricity is needed, as well as pump maintenance.

(See Appendices 8.2, 8.3, 8.4)

5.2   Treatment Wetlands

Treatment wetlands are lined treatment systems which support wetland vegetation in typically a gravel or soil medium such that as the effluent passes through the system it is treated by a combination of physical, chemical and biological mechanisms. Treatment wetlands are generally categorised as soil based or gravel based systems, the gravel systems being further subdivided into Vertical Flow and Horizontal Flow systems.

Reed Beds and Constructed Wetlands are two terms often used interchangeably to describe treatment wetlands. In Ireland the term Constructed Wetland is usually used to describe a soil based marsh system or Free Water Surface (FWS) wetland, in which the wastewater flows over the soil substrate. A Reed Bed is usually a gravel-based system in which the wastewater flows vertically or horizontally through the gravel substrate. These are often termed Horizontal Sub Surface Flow (HSSF) and Vertical Flow (VF) Reed Beds.

Ponds are sometimes incorporated into the design of both soil based constructed wetland systems and gravel reed beds for aesthetic value and treatment function. Due to their depth, they offer an increase in retention time without a corresponding increase in surface area. However it has been my practice over the past ten years or so to omit ponds from domestic system designs for safety reasons unless the homeowner specifically requests them. A pond may be included as an add-on to a constructed wetland or gravel reed bed where desired, but a high degree of pre-treatment in the wetland is required first to avoid excessive algal growth, and careful fencing is recommended for safety.

Treatment wetlands lend themselves to hybridisation, combining mechanical treatment systems or packaged filter units with treatment wetlands of various types, or different treatment wetland types mixed and matched at different treatment stages to achieve the overall treatment aims. An extension of this is the Living Machines system developed by John Todd in the US, but used around the world. This system uses a combination of mechanical aeration, media filtration and wetland planting to achieve the desired treatment aims in an aesthetically designed packaged system, usually housed in a greenhouse to keep temperatures up while maximising light. While you can mix and match to some extent with your treatment wetland options, bear in mind that different systems have different design criteria, so make sure that each component of your overall system is well designed and will work effectively.

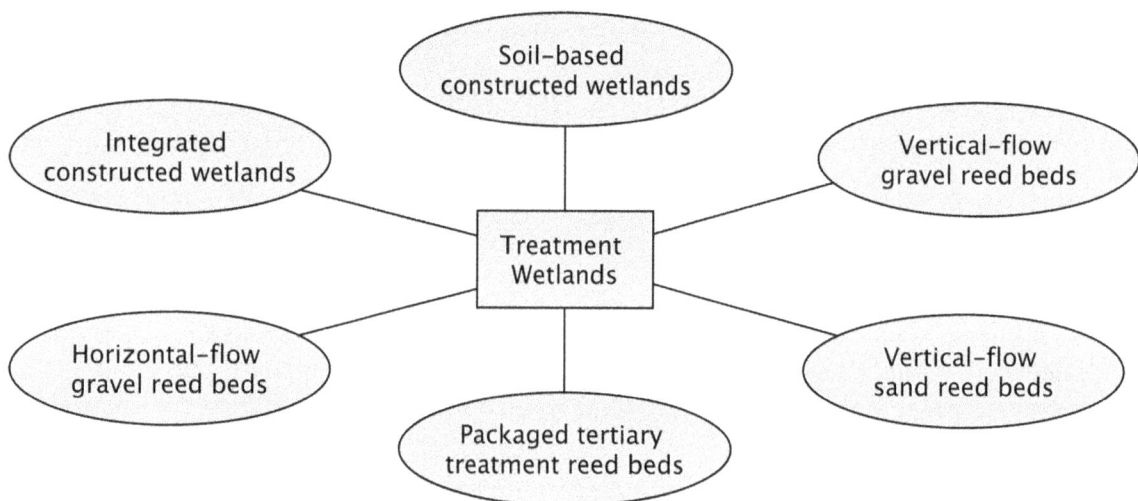

System layout within the site

Typically a septic tank precedes the treatment wetland, and should be desludged annually as usual. The wetland or reed bed can function effectively without electricity or pumps if there is a slope from the house to the septic tank, then to the treatment wetland and to the final percolation area.

If you have a small site and need a smaller treatment wetland, then this may be used following secondary aeration in a mechanical treatment unit. This can enable treatment wetlands to be used on relatively small sites where additional filtration of the effluent is needed or desired. VF sand reed beds and packaged reed beds are typically for tertiary treatment only.

Most treatment wetlands should be fenced after construction to keep children, livestock and the general public away from either open water or effluent at the gravel surface. Some gravel reed bed designs do not have exposed effluent and thus may not require fencing.

Pros and Cons compared to other treatment options

Pros:
- Can have comparatively low construction and running costs.
- Can have zero electricity inputs where pumping of effluent is not needed.
- Robust systems, tolerant to variable or seasonal loading rates.
- Can be designed to achieve excellent reductions of biochemical oxygen demand (BOD) and suspended solids (SS) from septic tank effluent where needed.
- Potential for good removal of a wide range of pollutants.
- Secondary benefits in terms of potential wildlife habitat enhancement and visual aesthetics.
- Versatile systems that may be used in conjunction with old or overloaded systems to achieve high discharge standards and legal compliance without excessive costs.

Cons:
- Treatment wetlands can take up too much land area for some sites.
- Where effluent or open water are exposed, these need to be fenced securely from children and livestock.
- As with any wetland habitat they can attract insects. This is seen by some as an advantage (e.g. dragonflies, damselflies and butterflies), however excessive proliferation of some insect life can potentially be a nuisance.
- Potential for odour generation, depending on the degree and type of preliminary treatment and type of treatment wetland. This is usually exacerbated by strong chemicals such as bleaches, which reduce the effectiveness of septic tank bacteria. Can be minimised by choosing environmentally safe cleaning products and by carefully siting the treatment wetland.
- Gravel reed beds require a septic tank maintenance regime to ensure that solids are kept out. Without this they are prone to clogging and may need full replacement of all gravel and plants.
- Sand reed beds require very careful pre-treatment to avoid clogging and malfunction.

Maintenance requirements for treatment wetlands

All treatment wetlands require some maintenance to ensure that they remain free of septic tank or treatment system sludges and some routine inspection to check that they are functioning, as they should, as follows:

- Ensure that the septic tank or secondary treatment system is kept desludged as per the supplier's recommendations. This will avoid carry-over of solids to the treatment wetland.
- Keep the system moist during dry conditions in the first season and possibly beyond, to keep plants healthy.
- Eliminate or minimise bleach and other toxic materials from the waste stream to maximise bacterial activity and the effectiveness of the whole sewage treatment system, including the treatment wetland. Paints, solvents, hydrocarbons and such materials should be kept out of the sewer system at all times.
- Water levels should be checked periodically to ensure maximum efficiency of the system.
- Inlet and outlet pipes should be checked periodically to ensure a free flow of liquid both into and out of the treatment wetland.
- Trees that may germinate and grow naturally within 5m of the system should generally be removed to minimise risk of damage to the liner.

- Some gravel reed bed designs call for removal of the surface vegetation after the end of the growing season.
- Vertical-flow reed beds require regular inspection and occasional maintenance to ensure that the surface does not become caked with sludges and block.

Soil-based Constructed Wetlands

Soil-based constructed wetlands or free water surface (FWS) wetlands are purpose-built wetlands, which are specially designed for wastewater treatment, storm water management or habitat enhancement. A carefully chosen selection of plants and a specially designed wetland area provide the right biological environment for cleansing and reoxygenating the water. These systems are modelled on natural wetlands, but are designed to achieve optimum treatment efficiencies. Natural wetlands and their plant communities have evolved to thrive on nutrient rich, silted waters. They have even been used inadvertently for sewage treatment since the first towns and villages channelled their waste into them.

Soil-based constructed wetlands can be the most cost effective treatment wetland option where marl clay conditions are present on site to negate the need for plastic lining. They are also the most robust of the treatment wetland types, since any accumulation of sludge or sediments can be removed by excavation without disturbing any gravel or liner materials.

Constructed wetland systems can be built on any soil type if they are lined with plastic to hold the effluent within the system. Even on relatively heavy clay soils a plastic liner is sometimes recommended if the clay impermeability is variable and there is a risk that groundwater ingress or effluent leaks will occur.

On existing sites, if the soil type is too heavy for a percolation area, a constructed wetland can be used to polish the effluent (from either septic tanks or mechanical treatment systems) prior to discharge to a stream or river. Sometimes heavy soil will be unsuitable for receiving a percolation discharge directly from the septic tank but will adequately percolate polished effluent.

For one-off houses, constructed wetlands can be used to effectively treat septic tank effluent, secondary treated effluent or grey water only, as appropriate. They can make attractive wetland areas both aesthetically and for wildlife.

(See Appendix 7.3)

System layout within the site

Constructed wetland systems are open marsh (or free water surface, FWS) systems, which receive effluent either from a septic tank or mechanical treatment unit prior to discharge.

A constructed wetland for a single three-bedroom dwelling is typically about 100m^2 internally and should be positioned c.50m from the house. An acre or more is the ideal size for a site that will comfortably accommodate a constructed wetland system. Below this size the open nature of the system may become a limiting factor in the overall garden design layout, which often also needs to incorporate play areas for children, fruit and vegetable gardening areas, lawn and ornamental areas, as well as driveways, and parking areas. A one acre site allows space for all of these elements of the garden plus space for the septic tank, constructed wetland and percolation area, all of which need to observe specified legal separation distances from the house, adjacent properties, wells and watercourses.

For smaller sites a gravel reed bed system or a mechanical aeration system may be more appropriate than a soil-based constructed wetland. These would be smaller in area and are generally covered by gravel, minimising the need for large separation distances from the house and adjacent properties.

(See Appendices 8.3, 8.5, 8.6)

Pros and Cons compared to other treatment wetlands

Pros:
- The open nature of constructed wetlands means that they are very resilient to sludge overloading and hydraulic shock loading (i.e. sudden overloads of effluent).
- Where combined sewers are present, receiving storm water from roof surfaces as well as grey and black water, constructed wetlands can be designed to accommodate these rainfall-dependent flow patterns and still produce reliable effluent quality.
- They are potentially the most cost effective systems available where subsoil conditions negate plastic lining. In such sites, full secondary and tertiary effluent treatment can be achieved for not much more cost than excavation and planting, following the septic tank.
- Hand in hand with cost effectiveness is energy and resource efficiency. Clay lined constructed wetlands can be the lowest embodied energy input sewage treatment systems available (excluding dry toilets which don't even need the septic tank and willow facilities which pay back carbon in firewood for every year after construction).
- They are probably the best treatment wetlands for wildlife since they directly resemble marsh habitats, albeit nutrient enriched ones.

Cons:
- Their open nature means that they are more susceptible to odour generation.
- They contain open water to a depth of *c.*200mm, which can be a safety hazard unless suitable precautions are taken.
- Due to their open nature they can be unsuitable for small sites.
- From an EPA *Code of Practice* standpoint they require more land space than gravel reed beds, however recent findings (reported in Kadlec and Wallace, 2009) suggest that the effectiveness of soil and gravel systems may be about the same.

Integrated Constructed Wetlands

Integrated Constructed Wetlands (ICWs) are essentially soil-based or free water surface (FWS) constructed wetland systems that take greater cognisance of landscape fit and habitat enhancement. The design basis for a soil based wetland and an ICW system is essentially the same, except that for ICW systems the plant species diversity is generally greater. Clay lined ICW systems can rely on infiltration and evapotranspiration to have zero discharge during the growing season.

The *Integrated Constructed Wetlands Guidance Document for farmyard soiled water and domestic wastewater applications* was produced by the Department of the Environment, Heritage and Local Government in 2010 and provides a very good overview on system implementation. The ICW Guidance Document also provides guidance on existing buildings where storm drains are already included in the sewer network, thus providing a documented method of dealing with such situations (namely doubling the system size) rather than having to dig up and replace sewers where this is undesirable.

Pros and Cons compared to other soil-based constructed wetlands

Pros:

- Like constructed wetland systems, integrated constructed wetland systems are an open marsh system and as such provide good habitat for wetland wildlife, particularly birdlife.
- Their design has a greater diversity of plant species than standard soil based wetlands, thus providing additional wildlife interest.
- Landscape fit is a specific design requirement, which isn't always prioritised in other constructed wetland designs.
- Zero discharge can be achieved at certain times of year on certain soils.

Cons:

- The large diversity of plant species can increase the cost of the planting work without providing significant additional treatment benefit.
- The requirement for landscape fit can increase the total land-take, and make the system more costly than may otherwise be the case.

Horizontal-flow Gravel Reed Beds

Horizontal-flow gravel reed beds or subsurface flow (HSSF) reed beds are relatively shallow, gravel filled basins planted with a selection of wetland plants which provide aeration and filtration to the effluent passing beneath the surface of the gravel.

Horizontal flow reed beds are generally smaller than constructed wetlands, however Kadlec and Wallace (2009) suggest that the efficiencies may actually be similar. The fact that the effluent is covered in gravel lends them to smaller sites than those required by soil-based constructed wetlands.

(See Appendix 7.4)

System layout within the site

Gravel reed beds are typically 30-50m² for a standard 3-bedroom house for secondary treatment. Positioning from the house should be about 30-50m from the dwelling, rather than the general EPA recommendation of 10m between a dwelling and any treatment system.

(See Appendices 8.4, 8.7, 8.8, 8.9)

Pros and Cons for HSSF reed beds compared to other treatment wetlands

Pros:

- Generally smaller than soil-based constructed wetland systems.

- Since the effluent is covered in gravel, there is no open water hazard and odour generation can be reduced.
- Water levels are more fixed in HSSF gravel reed beds, which means that the design can be selected to work effectively on sites where minimal fall is tolerated.

Cons:
- Generally more expensive than soil-based constructed wetland systems for single house applications, due to the need for gravel and more robust liner requirements.
- Maintenance input generally higher than for constructed wetlands.
- Septic tank maintenance needs to be much more strictly adhered to than for soil-based constructed wetlands. Considerably more troublesome to desludge than constructed wetlands in the event of sludge loading.
- Requires occasional removal and replacement of substrate and plants when the system clogs up. This may be 10-20 years in a well functioning secondary treatment reed bed system and potentially a lot longer for a tertiary treatment reed bed.
- There may be exposed effluent at the inlet section before it drops into the gravel medium, but this can be designed out if necessary.

Vertical-flow Gravel Reed Beds

Vertical-flow reed beds are essentially an extension of traditional trickling filters, with plants added for additional nutrient uptake and to assist with drainage. Both systems aerate the wastewater and provide a high degree of surface to air contact for biological filtration of the effluent. For those readers without a background in wastewater engineering, a scum layer of bacteria accumulates on the surface of each piece of gravel, and it is these bacteria that provide the majority of the treatment within vertical flow reed bed systems.

This system is an open chamber or tank, in which layers of progressively smaller grades of gravel (from the bottom up) and a top layer of sand overlay a drainage pipe network at the base of the tank. The effluent is typically pump-fed into the top of the tank through a network of pipes which allow the effluent to spread over the entire surface of the bed in each pump cycle, before filtering down through the gravel.

(See Appendix 7.5)

System layout within the site

The VF reed bed will typically receive effluent from either a septic tank or from a mechanical secondary treatment system. The final effluent from the vertical flow reed bed typically goes to a horizontal-flow reed bed for further treatment prior to percolation or surface discharge. Alternatively a polishing filter as per the EPA *Code of Practice* or a standard percolation area may be used instead of a HSSF reed bed for the second stage of treatment. Note that where a VF reed bed is not followed by a HSSF reed bed, a deeper bed depth is recommended than the EPA *Code of Practice* specification.

Sizing given in the EPA *Code of Practice* is c.15m^2 for secondary treatment for a 3-bedroom house (5pe) for secondary treatment (i.e. directly from a septic tank).

(See Appendices 8.4, 8.7, 8.8, 8.9)

Pros and Cons compared to other treatment wetlands

Pros:

- Due to the pumped nature of the treatment process, vertical-flow reed beds can achieve greater oxygenation of the effluent within a smaller surface area. This can be useful for smaller sites.
- The oxygenation also assists in the reduction of ammonia concentrations in the wastewater. Ammonia is one of the more malodorous components of sewage effluent, so this can be helpful where a low odour system is important. That said, care with the cleaning products such as detergents and bleaches in the house may have just as significant an impact on odour as the type of system selected. The healthier the septic tank bacteria, the better.

Cons:

- A pump is typically needed to get the effluent to the spreading pipes within the system, and sometimes to draw the effluent from the base of the system to the next stage of treatment.
- Typically these systems are used in conjunction with a horizontal-flow reed bed, so can sometimes be seen as an additional use of space and cost, rather than a substitute for a larger system.
- The incoming liquid needs to be of a sufficiently high quality – specifically, it needs to be low in suspended solids – to avoid clogging the gravel or sand layer at the top of the reed bed.

Vertical-flow Sand Reed Beds

Vertical flow sand reed beds are a hybrid between sand filters and reed beds, used typically for tertiary treatment of pre-treated effluent. The sand provides a finer filter than a VF gravel system, almost like a sieve, screening out any remaining debris that may come from the preceding treatment system. The presence of the wetland plants in a vertical flow sand reed bed provides additional treatment around the root zone as well as uptake of nitrates and phosphates during the growing season and also helps to keep the sand medium open for to avoid premature clogging of the system.

System layout within the site

Like the VF gravel reed bed, this system is an open chamber or tank with a pumped feed to allow the effluent to spread over the entire surface of the sand in each pump cycle.

Sand filled VF reed beds are not as common in Ireland as VF gravel reed beds (which may be topped with sand). The potential for system clogging is higher in a sand reed bed, and these should only be used where maintenance of the preceding treatment system is going to be prioritised by the homeowner.

The VF sand reed bed will receive effluent from a septic tank or more typically from a mechanical secondary treatment system. Sizing given in the EPA *Code of Practice* is 25-30m² for a 3-bedroom house for secondary treatment. However given the tendency for sand to clog if suspended solids levels are at all elevated, I would encourage the use of gravel as a medium for secondary treatment applications. Refer to section 5.9 on tertiary treatment for sizing of VF sand reed beds for effluent polishing.

Pros and Cons compared to VF gravel reed beds

Pros:
- Sand will give a greater surface area per m³ than gravel as a filter substrate. This means that the bacterial scum that accumulates within the system has the potential to have a much larger treatment area and provide better effluent polishing than gravel.
- Effluent will move more slowly through the sand substrate, providing extended treatment time as compared with gravel.

Cons:
- System clogging is much more likely, so greater pre-treatment is important to avoid system failure or excessive maintenance input.
- Greater inspection frequency is also recommended.
- The system size recommendations in the *Code of Practice* are greater than for gravel systems.

Packaged Tertiary Treatment Reed Beds

Tertiary treatment reed beds are sometimes supplied as an add-on to provide additional filtration after mechanical treatment units. These are small horizontal flow gravel reed bed systems that offer biological filtration of the final effluent by bacteria on the gravel surface and around the plant root zone and also provide uptake of nitrates and phosphates during the growing season.

System layout within the site

These units are typically sold as a pair of small rigid plastic tanks, approximately 1m x 2.5m each. They are installed as part of the secondary treatment system installation and filled with pea gravel or other suitable gravel grade and planted with either common reed (*Phragmites australis*) or a small selection of appropriate wetland plants such as *Phragmites*, bulrush (*Typha latifolia*), yellow flag (*Iris pseudacorus)* and water mint (*mentha aquatica*).

The final effluent from the packaged tertiary treatment reed bed goes either to a polishing filter as per the EPA *Code of Practice*, or directly to a surface discharge as the site conditions dictate.

Pros and Cons compared to other treatment wetlands

Pros:
- Packaged systems are smaller in total surface area, and thus fit more easily into small sites.
- They are more easily installed, since the rigid plastic unit arrives on site and simply requires filling and planting to set it up – in addition to standard installation measures such as excavation, plumbing, backfilling and landscaping.

Cons:
- Although they meet the EPA minimum size for 5pe, they are very small in terms of treatment area, so it is all the more important that the secondary treatment system achieves its design objectives on a consistent basis in order to provide adequate protection of wells, groundwater and/or watercourses. It is by no means unheard of for secondary treatment systems (whether these be mechanical treatment units or treatment wetlands) to fail in meeting consistent discharge limits.

- In common with other gravel reed beds, suspended solids loads will accumulate at the inlet section and cause clogging of the system over time, although this timeframe will be extended for systems following well functioning and well maintained secondary treatment systems.

5.3   Mechanical Treatment Units

Mechanical treatment units became increasingly popular in Ireland in the late 1990s and early 2000s when the acceleration in house building coincided with an increasing awareness of the need to deal responsibly with domestic effluent. Mechanical treatment units oxygenate settled sewage effluent by use of air diffusers, pumps or other mechanical methods to bring about a reduction in BOD, suspended solids and ammonia concentrations. This process of aeration of effluent followed by settlement and removal of the resulting biosolids is referred to secondary treatment.

I want to flag here that mechanical treatment units aren't my thing – my knowledge of them is relatively patchy and my enthusiasm somewhat limited. Their on-going electricity requirements without a return of habitat value, willow wood or biogas is something that isn't environmentally sustainable enough for me. Given evolving climate change policies and peak oil, who knows what energy issues we may face in the coming decades, and whether a reliable electricity supply is something we can count on to keep our waterways and groundwater resources clean.

That said, if you choose to disregard peak oil and climate change, or if other factors come into play such as limitations on site size, then mechanical treatment units should certainly be considered. They are very good at improving the quality of sewage effluent within a compact, tidy, covered unit, and are currently a common choice for people who need to achieve secondary treatment standard.

Bearing in mind this personal bias, the different mechanical treatment system types are outlined in this section to help you choose the system most suited to your site and your personal preferences. With some exceptions, many treatment systems are quite similar in their mode of operation, so treat this section more as an overview of options and do an internet search for the different sewage treatment system suppliers to find out more about specific products.

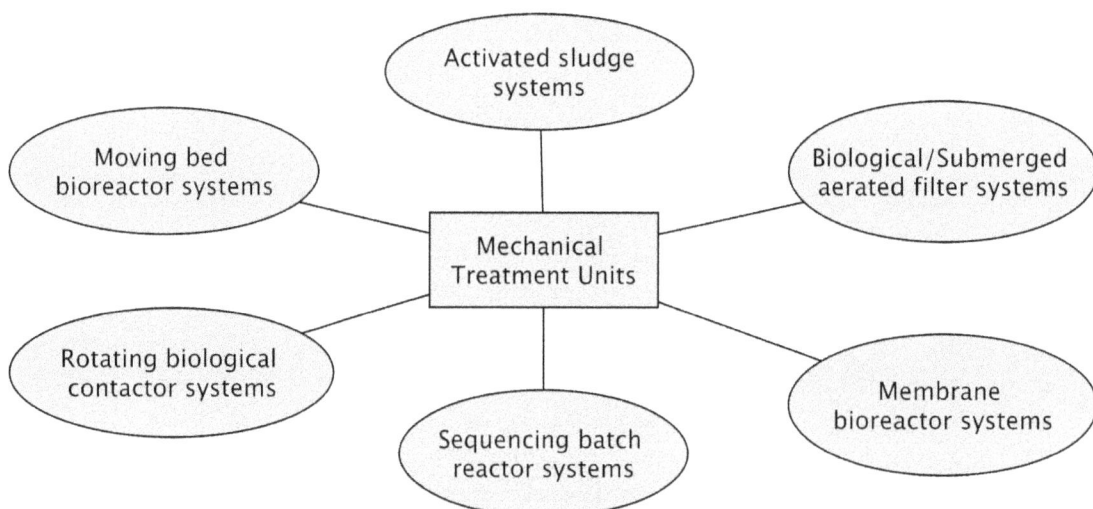

System layout within the site

Mechanical treatment units can fit anywhere on a site that a septic tank can fit, so they are the most versatile secondary treatment option in terms of positioning and layout. The secondary treatment that they offer enables the percolation area or polishing filter following this stage of treatment to be smaller than the equivalent area following a septic tank.

(See Appendices 8.2, 8.5, 8.9)

Pros and Cons compared to other treatment options

Pros:
- Mechanical treatment units (and packaged media filter units) have the advantage that they can fit into a smaller site area than most of the other secondary treatment systems available.
- Maintenance can generally be handed over to a maintenance operator, so the input of the homeowner is generally minimal.

Cons:
- The drawback of mechanical units is that their electricity use can be considerable. A 100W air blower going 24 hours a day for 365 days per year will use 876kW of electricity, costing nearly €170/year (at 19.31c/kWh, July 2014 prices) which will produce $c.$460kg CO_2/yr.; in addition to standard electricity costs and carbon footprint for the home. Note that not all mechanical units require 24/7 electricity, see RBC units described below.
- The flip side of minimal maintenance inputs on the part of the homeowners is that the annual maintenance cost and the costs of part replacement and repair is typically more than for non-mechanical systems. Some homeowners specifically like to have more input into the running and maintenance of their treatment system, and the nature of the mechanical unit is that it tends to require specialist maintenance.

Maintenance requirements for mechanical treatment units

The maintenance requirements of mechanical treatment units are relatively specialised, so homeowners are generally encouraged or required by the county council to enter into a maintenance contract with their supplier to carry out this work. In terms of homeowner input, the following maintenance factors are recommended:

- Ensure that you contact your maintenance contractor annually or as needed to carry out the necessary works.
- Ensure that the system is kept desludged as necessary if your maintenance contractor does not carry out this as part of their brief.
- In order to maximise the health and effectiveness of the bacteria within your system it is recommended that you minimise the use of bleaches, and that the cleaners and household chemicals are as septic tank friendly as possible.
- Keep the system plugged in and turned on. This may sound like a fairly self evident requirement, but nonetheless it isn't unheard of that systems have been installed but never actually used for anything other than an expensive and inefficient septic tank.
- Carry out routine inspections on the alarm and component parts – even just listening to hear that the blower is on and the pump, if present, is kicking in every so often to remove effluent to the polishing filter as needed.
- Carry out routine inspections on the polishing filter area as needed.

Conventional Activated Sludge (CAS) Systems

The activated sludge system is perhaps the most straightforward of the mechanical treatment systems available for domestic applications. Like all secondary treatment systems the process basically involves a primary settlement stage, akin to a septic tank within the treatment unit. Following this, the secondary treatment stage in the activated sludge process involves direct aeration of the primary settled effluent by means of air diffusers at the base of the tank. In this process the microorganisms active in this stage of treatment reduce the organic contaminants in the wastewater. After this aeration tank, a clarification chamber is used to allow the microorganisms in the treated effluent to settle out prior to discharge of the liquid. Discharge in this case may refer to a percolation area, polishing filter or surface discharge.

Pros and Cons compared to other mechanical treatment units

Pros:
- Generally less maintenance requirements than other mechanical systems due to fewer moving parts and straightforward operation.
- Can achieve a greater level of BOD removal than other mechanical treatment systems. (Dubber and Gill, 2012.)

Cons:
- May be prone to hydraulic surges and flushing of active bacteria from the system.
- Can have comparatively poorer suspended solids removal. This type of system is called a 'suspended growth system' in which the bacterial growth is suspended in the liquid – as opposed to 'attached growth systems', in which bacteria grow on the plastic media surface within the tank. Suspended growth systems can suffer from poor sludge settlement as compared with attached growth systems.

Sequencing Batch Reactor (SBR) systems

The SBR system operates like a conventional activated sludge system except that the different stages of aeration are carried out in a batch rather than a continuous stream. In municipal SBR systems, two tanks are used in alternating cycles. The first tank fills; is aerated during the treatment phase; sludge and liquid separate during the settlement phase; is emptied of sludge and liquid during the drawdown phase. Then the cycle takes place in the second tank, which was filling during that time. In domestic SBR units there are usually two tanks used in series rather than parallel. Tank 1 fills while tank 2 has the aeration, settlement and drawdown phases. Then the contents of tank 1 are pumped into tank 2 and the cycle continues.

Pros and Cons compared to other mechanical treatment units

Pros:
- Due to the rest-phase of the SBR system, they have slightly lower electricity requirements than CAS and SAF systems. (Dubber and Gill, 2012.)
- Good at removing total nitrogen due to the alternating cycles of aeration and resting.

Cons:
- Being suspended growth systems, sludge settlement can sometimes be suboptimal. (How many Ss' is that?)

Biological/Submerged Aerated Filter (BAF/SAF) Systems

Biological or Submerged Aerated Filter (BAF/SAF) systems work by blowing air up through a plastic media 'filter' or framework that supports the growth of a bacterial film. Typically the media filter is fixed in position, or where it is in suspension within the liquid it is referred to as a Moving Bed Bioreactor (MBBR). It is the film of microorganisms that grow on the surface of the plastic filter media that treat the waste by consuming it's food value (or pollution value) and converting it to bacteria mass, which settle out of suspension and slough off the plastic structure to settle in the base of the unit or a following settling chamber. It is a relatively straightforward system, providing a hybrid between the air pumping employed in the activated sludge system with the media structure for bacterial growth of the rotating biological contactor and fixed film filter approaches.

Pros and Cons compared to other mechanical treatment units

Pros:
- Attached growth systems settle solids out well, and are thus very robust compared to suspended growth systems such as CAS and SBR systems.

Cons:
- Due to the requirement for 24/7 oxygenation, this system, in common with CAS systems, has a high electricity demand, with associated carbon footprint and costs. (Dubber and Gill, 2012.)
- MBBR type systems are more expensive to purchase than most other mechanical treatment units. (Dubber and Gill, 2012.)

Rotating Biological Contactor (RBC) Systems

All mechanical treatment systems require oxygenation of the effluent to achieve aerobic 'secondary' treatment. In this way the plentiful supply of oxygen and the abundant food value of the contaminants in the effluent provide optimum conditions for bacteria to thrive and consume this 'food', thus cleaning up the water. Like BAF/SAF systems, RBCs rely upon a rigid medium on which the bacteria grow, but unlike the former, the medium or 'biological contactor' rotates through the effluent, alternating between being alternately submerged and then exposed to air as it does so. In this way, bacteria thrive on the oxygen supplied by direct contact with the air on the rotating framework and the food value supplied by the contact with wastewater in the tank. Thus as liquid passes through the RBC unit, the food value of the effluent is consumed and in this way, the wastewater quality is improved.

Pros and Cons compared to other mechanical treatment units

Pros:
- Generally uses approximately 50% less electricity than systems with continuous pumped aeration such as CAS and SAF units.
- Since this is an attached growth system, sludge settlement is generally more reliable than CAS and SBR systems, which are suspended growth systems.

Cons:
- Has the potential to require greater maintenance due to more moving parts.
- More moving parts available for potential mechanical faults.

Membrane Bioreactor (MBR) Systems

A membrane bioreactor is similar to the activated sludge system in that there is an aeration zone in which air is bubbled through the effluent. Following this zone, the effluent is filtered through a fine membrane rather than settled out like in an activated sludge system. This filtration stage affords a treatment standard typically greater than a gravity settlement process, to enable a higher standard of effluent to be reached.

Pros and Cons compared to other mechanical treatment units

Pros:
- Provides the highest treatment standard compared with other mechanical treatment units.

Cons:
- By virtue of the system design, this system requires greater maintenance input to keep the membrane free of clogging.
- Most costly domestic scale packaged treatment system – both in terms of purchase price and electricity costs. (Dubber and Gill, 2012.)

5.4   Packaged Filter Media Units

Packaged filter media units entered the market place at about the same time as the mechanical treatment units, designed to carry out essentially the same task. For normal domestic scale applications these filter units are secondary treatment systems, designed to reduce the concentration of BOD and suspended solids in the final effluent.

In these units, effluent is pumped or gravity fed from the septic tank or settling stage of the treatment process to a distribution piping network above the filter media. The pressure feed itself in pumped systems introduces air for the biological treatment process. The effluent percolates down through the filter medium where it is exposed to the microbial flora that builds up on the surface of the media layer. The high surface area of the filter media provides a high degree of surface to effluent contact for effective biological treatment. The final effluent is collected in the base of the system and is either piped to a percolation area or polishing filter, or percolated directly into the ground beneath the unit via a gravel polishing filter.

The media substrate may be anything that has a relatively robust open structure to avoid rapid clogging; has a relatively long, stable life under conditions of regular saturation; and has a high surface area to maximise the surface to air contact for the microbial flora that actually carries out the biological treatment. Soil, sand and gravel are the traditional filter media units and these are the primary media used for polishing filters and vertical flow reed beds. However, these materials are heavy to transport and relatively difficult to replace in the confines of a packaged unit. Lighter examples that are used in packaged units include such media materials as rock wool, plastic, textile and coconut fibre.

The primary difference between packaged media filters and either polishing filters or vertical-flow reed beds is that they are standardised off-the-shelf units that be can dropped into position on-site rather than constructed *in situ*.

```
        ┌─────────────────────┐
        │   Pump–fed filter   │
        │   media systems     │
        └─────────────────────┘
                    │
         ┌────────────────────┐
         │  Packaged Filter   │
         │    Media Units     │
         └────────────────────┘
                        │
              ┌─────────────────────┐
              │  Gravity–fed filter │
              │   media systems     │
              └─────────────────────┘
```

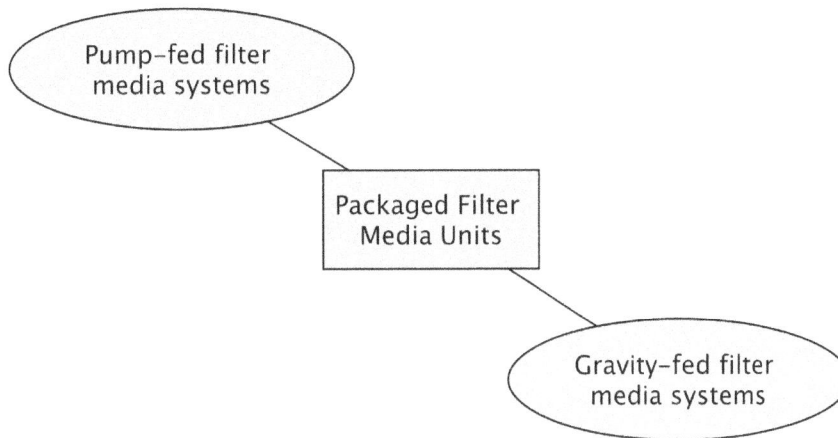

System layout within the site

These units are a similar size to mechanical aeration units, and as such can fit easily into most gardens. They are typically used in tandem with a septic tank, so there needs to be space to accommodate both the tank and the treatment unit. However the polishing filter for final effluent disposal may be located beneath the unit in sites of suitable ground conditions, so this can offer an overall space saving.

Pros and Cons compared to other treatment options

Pros:
- The filter media can offer good BOD and suspended solids reductions.
- Lowest electricity consumption of packaged treatment options available – even for pumped models – offering both cost savings and carbon savings over all mechanical treatment units.
- Considerably smaller footprint area than treatment wetlands or willow facilities.
- Maintenance can be handed over to a specialised contractor for ease of use for the homeowner.
- May be used in conjunction with an existing septic tank as an add-on filter where needed.

Cons:
- Pumping, where needed, requires an electrical input.
- Media requires replacement when saturated with scum or sludge – This varies with system type and the degree of maintenance of your septic tank, so check with the suppliers before selecting a system and keep you tank well maintained.
- Maintenance generally requires specialist input.

Maintenance requirements for packaged filter media units

There is relatively little maintenance input required of the homeowner for these systems. However as with all treatment systems, some routine maintenance is required to keep the system functioning effectively.

- Ensure that your system is checked annually or as needed by a specialist contractor and that any necessary maintenance work is carried out promptly.
- Ensure that your septic tank is kept desludged as necessary to avoid solids carrying over to your spreading system and clogging the filter medium prematurely.

- Replace your filter medium as necessary to avoid having the system block up and malfunction or overflow.
- In order to maximise the effectiveness of the microorganisms within the system it is recommended that you minimise the use of bleaches and that the cleaners and household chemicals are as septic tank friendly as possible.
- Carry out routine inspections on the pump and alarm, if present, to ensure that these are working effectively.
- Carry out routine inspections on the area around the polishing filter to ensure that there is no surface ponding or other indicator of malfunction.

Pump-fed Filter Media Systems

These systems function by pumping the effluent from the septic tank to a spreading system above the filter medium within contained plastic units. The combination of aeration due to pressurised application of the effluent and filtration by the medium itself and the bacterial film that develops on its surface provide for the reduction of BOD and suspended solids. The most common media type for pumped filter systems is coconut fibre.

Pros and Cons compared to gravity-fed filter media units

Pros:
- Pumping can help to achieve a good spread of effluent on the media surface.
- Dosing of effluent allows cyclical wetting and drying of the media, which can benefit treatment effectiveness.
- Pumping can allow the units to be raised above ground, sitting directly over the percolation area, which can have space savings over gravity systems.

Cons:
- Typically uses coconut fibre, which has to travel a great distance to get to Ireland, and with that is an associated carbon footprint.

Gravity-fed Filter Media Systems

Gravity-fed packaged filter media systems are becoming available as the need for zero electricity treatment systems is being recognised. Gravity-fed systems typically have a low outlet pipe, so on flat sites a pump may still be needed to convey effluent away from the system after treatment or to lift it to a suitable elevation for discharge via percolation.

Pros and Cons compared to pump-fed filter media units

Pros:
- Where topography permits, they can work without any electricity inputs.
- Some use renewable, low embodied energy substrates such as coconut fibre.

Cons:
- Low outlet pipe invert, requiring suitable site topography or pumping of effluent up to discharge level.
- Some use high-embodied energy substrates such as rock wool.

5.5   Willow Treatment Systems

Willows have become increasingly popular as a final component of the treatment chain because they are excellent at mopping up excess nitrates and phosphates. Willow systems can be used to soak up all liquid if carefully designed and built.

There are a number of distinct ways in which willows are used in dealing with domestic wastewater. Three methods described here are willow treatment stems for filtering water *en route* to discharge; for improving the soakage and nutrient uptake around percolation areas and polishing filters; and for zero discharge facilities to dispose of all the effluent to air instead of to ground or to surface waters.

In all types of willow treatment systems the type of willow is an important factor in system performance. Where wastewater treatment is the primary objective, only certain willow cultivars should be used, specifically for high biomass production in nutrient enriched conditions. The most common cultivars are bred from the species *Salix viminalis*.

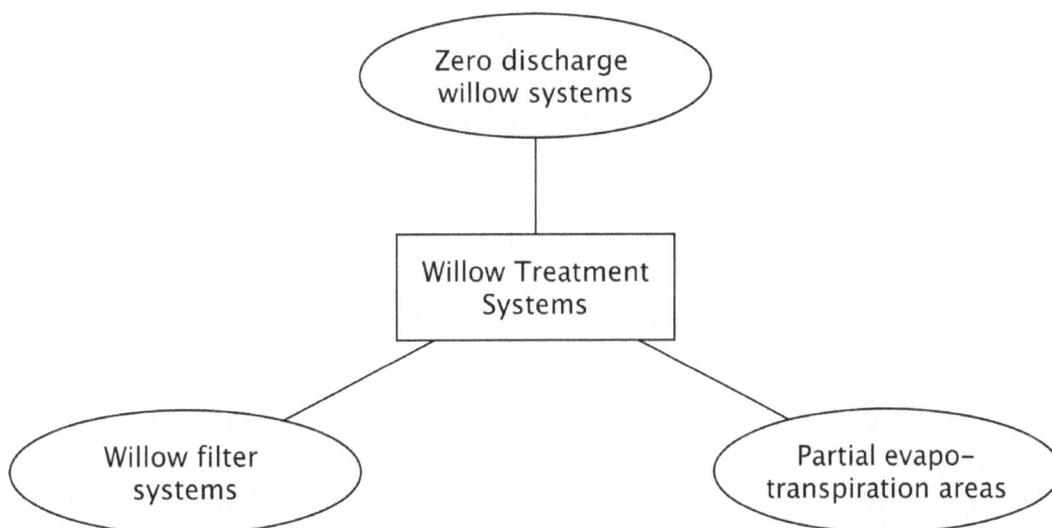

System layout within the site

By their very nature, willow treatment systems will be quite prominent site features, with some willow system types taking up more site area than others, as described further below. Willow trees also have extensive root systems that require judicious planting. In fully plastic lined systems such as zero discharge willow facilities root intrusion isn't a problem, but for partial evapotranspiration systems used around the outer perimeter of percolation areas to take up nutrients, additional space may be needed to avoid root intrusion into percolation pipes, or different percolation trench methods may be employed. Basically with all willow systems there is a preceding settlement tank or, under certain circumstances, a secondary treatment system.

Pros and Cons compared to other treatment options

Pros:
- Willow facilities have the distinct advantage over most other systems in that they can be designed for high uptake of nitrates and phosphates.

- Where gravity can be employed, willow filter systems can function without the need for pumps or air blowers, thus achieving a high effluent quality without electricity or specialist maintenance. However this is not the case for most zero discharge willow facility designs.
- The trees provide wildlife interest and landscaping interest.
- The rapid growth of willow trees in areas of high soil moisture and abundant nutrients leads to rapid biomass production which can be used for fuel or chipped for landscaping use.
- The high biomass production also takes up atmospheric carbon, which, if used instead of fossil energy, can actively lower your carbon footprint, as well as saving on fuel costs.
- Willow systems with a discharge have the advantage over other treatment systems in that the evapotranspiration can reduce or eliminate discharges to watercourses during dry periods in the growing season. This can be of particular benefit where the final discharge is to a small stream or open field drain. Historically, a 1:8 dilution factor (effluent flow: stream flow) has been required for surface discharges. Where this dilution rate is not available in summer months, a willow treatment system can help bring the discharge volumes down to acceptable levels when dilution is low.
- Zero discharge willow facilities have the further advantage over other treatment systems in that they do not have any effluent discharge at all, at any time of year. They can thus be used on sites of poor percolation characteristics instead of a surface discharge option.
- Willows can be added cost effectively and unobtrusively on existing sites where percolation areas are giving trouble. Note that care is needed to design the willow area such that it won't conflict with the percolation piping or other garden features.

Cons:
- Willow trees are large features within a garden and are not always suited to small sites.
- Willow systems can take up a considerable amount of space, and can be too large for sites below an acre unless they are incorporated into the overall garden design, which can be very effective – thus using the willows as a site boundary feature.
- The willows generally need to be coppiced on a rotation basis so that there is always an abundance of fresh stem growth for maximum effectiveness. This work may be undertaken by the homeowner or by a contractor as required.
- Zero discharge willow facility installation can be expensive compared to a septic tank and percolation area or constructed wetland.

Maintenance requirements for willow treatment systems

The beauty about willow treatment systems is that most of the maintenance falls well within the capabilities of most homeowners. For anybody who enjoys outdoor time or gardening, it can be a pleasant part of the gardening year. That said, there are specific maintenance factors to be considered, as follows:

- To maximise stem growth and development, coppicing on a rotation basis is important. This rotation is usually every three years, meaning that every year one third of the trees is cut back close to ground level and allowed to grow for the following three growing seasons, whereupon the cycle continues.
- Desludging of all preceding settlement tanks or treatment systems is necessary to prevent solids being carried over to the willow plantation.

- Weeding of the willow plantation is needed to allow the trees to grow at the most efficient rate possible.
- Regular checks on the system are necessary and remedial action carried out where required, particularly for the zero discharge willow facility.

Willow Filter Systems

Willow filter systems are very similar to horizontal flow reed beds or soil based constructed wetland systems, or vertical flow reed beds depending on the willow system design. These systems are typically fully lined or used on sites with impermeable clay. They are filled with soil or gravel and planted with willows. The effluent flows through the soil or gravel, past the roots of the trees, where the microbial activity of soil microorganisms and the nutrient uptake by the trees treat the effluent prior to discharge.

Pros and Cons compared to other willow treatment options

Pros:
- Willow filter systems have the advantage over zero discharge willow facilities in that they can be smaller and shallower since they do not have to achieve full storage/ evapotranspiration. In this regard they are less limited by site size and area less costly to construct.

Cons:
- Since an effluent discharge is still present for some or most of the year, these systems may still require a discharge licence where the council require it.
- There has been less research carried out using willow filter systems than treatment wetlands, so the base of experience is not yet present to provide comprehensive data or design guidance.

Partial Evapotranspiration Areas

Partial evapotranspiration areas are willow planted areas without the clear boundary of a plastic or even clay liner. They are designed to compliment percolation areas, and to absorb nitrates and phosphates from the groundwater around these systems as well as provide partial evapotranspiration of effluent. The evapotranspiration can dry ground that may otherwise be boggy or suffer from surface ponding. It can also, depending on the site and the system design, eliminate or reduce the volume of a surface discharge on sites of somewhat limited percolation.

Partial evapotranspiration areas are best planted in conjunction with pump-fed percolation systems, or located down gradient of a percolation system to provide groundwater improvement without risking root intrusion into the piping network. Another method is to use willows in conjunction with infiltration leachfield chambers, basically inverted half-pipes that allow effluent to flow over an essentially flat soil base. These allow the willows to reach the effluent without as much risk of clogging pipe pores. That said, willow roots can grow very quickly and the lifetime of any underground chamber is uncertain.

(See Appendices 8.8, 8.7, 8.10)

Pros and Cons compared to other willow treatment options

Pros:
- These areas are typically designed to fit in around the existing sewage treatment infrastructure and are thus very adaptable to site layout plans.
- They can provide a very useful add-on to existing raised percolation systems or between percolation areas and adjacent watercourses.
- Where they are used as a buffer zone (rather than treatment system *per se*) between the pollution source and watercourse, they can be very efficient on space since they needn't necessarily observe EPA minimum distances, and can essentially comprise only a row or two of willow trees along a boundary.
- They can provide good amelioration of existing drainage problems by providing a firm root structure just beneath the ground surface and an aesthetically pleasing approach to challenging site conditions such as surface ponding or damp lawn.

Cons:
- Willow roots are very invasive of percolation pipes and there is a risk of making a drainage problem worse if the design and implementation are not carefully carried out.
- It is relatively difficult and/or costly to quantify groundwater improvement, so the benefit of partial evapotranspiration areas can be difficult to assess. However, if your trees are growing very well it is an indication that there is plenty of nitrate and phosphate that needed to be taken out of the groundwater in the planted area.

Zero Discharge Willow Facilities

Zero discharge willow facilities are plastic lined, soil filled basins that rely upon the willows for 100% evapotranspiration of effluent. The facility is preceded by an over-sized septic tank for solids settlement. Effluent is pumped to a pressure distribution system, which disperses the effluent evenly along the full length of the facility, thus ensuring that all trees receive an equal nutrient load.

(See Appendix 7.6)

The willow planted basins are typically 6-8m in width, and are carefully designed to provide the appropriate storage volume and willow evapotranspiration rates for the household served. Too large and the system will take in excess rainfall. Too small and it won't hold all the volume required. Typical system lengths in the Irish context vary from 30-40m for most domestic applications. As with all willow systems, the trees are coppiced on a rotation basis to provide maximum evapotranspiration potential and maximum nutrient uptake.

(See Appendices 8.11, 8.12)

Pros and Cons compared to other willow treatment options

Pros:
- The most obvious advantage is that the system is designed to have no discharge at any time of year. Thus it can be used effectively on sites that fail the *Code of Practice* soil percolation and/or trial pit tests, or adjacent to sensitive watercourses without risk of causing pollution.

Cons:

- The main drawback of the zero discharge willow facility is that it is more costly to install than other willow systems due to the larger size and fully plastic lined construction.
- It is also larger in size than other willow systems, and thus may not suit every site.
- The requirement for a pump for the septic tank effluent feed requires the use of electricity. However this should be viewed in the context that the system is net beneficial from the perspective of carbon uptake over its lifetime if the coppice wood is used instead of oil heating.

5.6   Source Separation Systems

Over the past decade or so, various technologies have emerged to deal with some of the inherent problems associated with mixing high volumes of clean water with faecal biomass and nutrients in flush toilets to create sewage. The systems developed to meet this challenge are those that either fit with the sewer piping infrastructure or those that avoid the sewerage infrastructure completely. This section examines urine diversion options as well as faecal separation systems for use with a flush toilet infrastructure. Dry toilets are described in section 5.7, which aim to avoid flush water altogether.

Urine and faecal separators are not treatment systems as such, in that they do not provide primary, secondary and tertiary treatment in the same way as treatment wetlands, packaged treatment systems or polishing filters. However they can achieve the same general improvement in water quality by simply diverting the urine into a separate system and/or separating out the faeces and paper from the flush water for composting. Both urine and humanure (the term used to describe human manure) are useful agricultural fertilisers, particularly in the context of environmental sustainability. Indeed, how can we build a sustainable culture at all if we continue to consume nutrients from agricultural land and then refuse to return the same nutrients to close the cycle? Instead we currently make our nitrogenous fertilisers from natural gas and mine our phosphates from rock phosphate, both of which are finite resources. Meanwhile the high nitrate and phosphate value of our own urine and faeces are the very problem that we grapple with in wastewater engineering and environmental protection every day of the week. Something in there just doesn't add up...

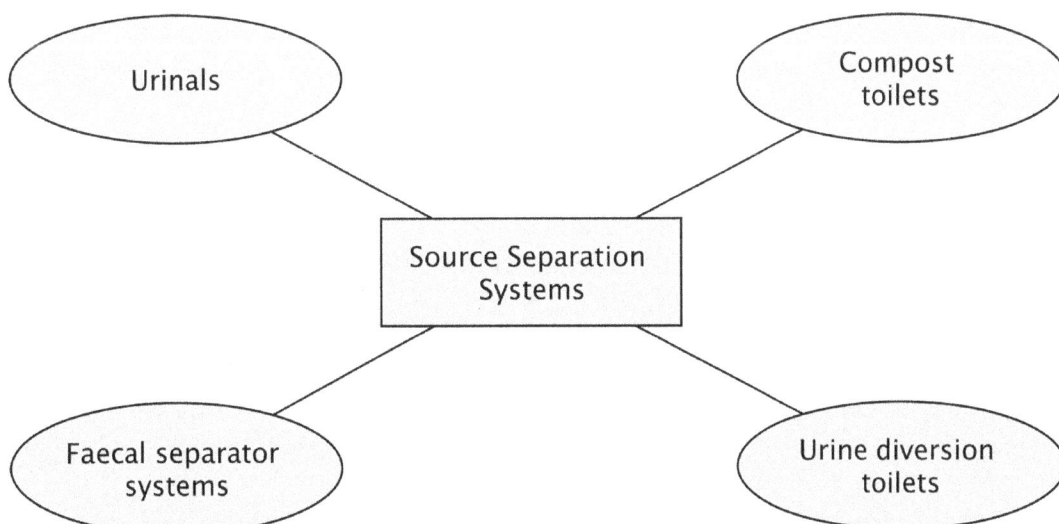

System layout within the site

In terms of site layout, urine diversion toilets and urinals do not take up any additional space other than the urine storage tank. Faecal separators need not take up much more space than a conventional septic tank. There is additional pipework required for urine diversion however, since the black water and 'yellow water' as it is termed in Scandinavia, need to be routed separately. A composting area is needed for maturing humanure.

Some form of polishing filter such as a treatment wetland, percolation area or constructed polishing filter is recommended prior to disposal of the flush water after urine diversion and faecal separation. If urine diversion is employed alone, then a septic tank or secondary treatment unit may be used to provide settlement or secondary treatment respectively. If only faecal separation is employed, then a treatment wetland or direct garden irrigation is recommended to remove some of the nitrates and phosphates in the resulting wastewater.

Pros and Cons of urine diversion and faecal separation compared to other forms of treatment

Pros:
- High removal of faecal microorganisms is possible with faecal separation.
- High removal of nutrients (>90% of nitrates and *c.*50% of phosphates) is possible with urine diversion alone.
- Nutrients are available for return to agriculture or horticulture under controlled conditions.
- Both can be zero energy systems with suitable site topography and/or house design.
- Since the homeowner can carry out all or most maintenance, this brings the annual running costs down, or eliminates them entirely.

Cons:
- Components relatively costly (compared to standard septic tank and percolation technology) due to being less easily available. However note that urine and faecal separation can be considerably less costly than a mechanical treatment system designed to achieve the same effluent quality.
- Requires more homeowner input and maintenance to keep the systems checked and emptied as necessary and to compost or otherwise mature the humanure from faecal separators.
- Some degree of storage time and space is needed for both urine and faecal compost; so sufficient site space needs to be available. Again, this storage area is small when compared to the surface area of a treatment wetland or willow system.
- Under current EU standards, return of humanure or human urine to certified organic lands is only possible with a specific exemption.

Maintenance requirements for faecal separators and urine diversion toilets

As with any treatment system, maintenance is crucial for the correct operation of faecal separation and urine diversion. Maintenance input for source separation tends to be greater than for standard flush toilet infrastructure. However, while it may take 5 years for a poorly maintained septic tank to clog up your percolation area, the feedback from separation and diversion systems is somewhat more rapid. This can be a positive or a negative depending on how much you want to ignore a problem, but there it is nonetheless.

- Empty the urine container as necessary and use in agriculture or the garden or

store for maturation if needed.
- Empty the faecal chamber as necessary and compost or mature the humanure for reuse in the garden or agriculture.
- Ensure that any other components of the overall wastewater management system are maintained and cared for as per the supplier's recommendations.

Urine Diversion Dual-flush Toilets

Urine diversion dual-flush toilets are essentially the same basic shape and design as conventional flush toilets, except that there is an additional mini-bowl to the front of the unit to divert urine to a separate outlet pipe. This urine is stored undiluted, or with the very small amount of flush water that is used to flush this mini-bowl. The urine is then used as a nitrogen and phosphorus rich fertiliser in agriculture or in the garden.

Urine contains well above 90% of the nitrogen content of sewage and about half of the phosphorus. Urine diversion toilets divert approximately 50-80% of the urine from a household, increasing with user education and care. In this way a significant proportion of the total nitrogen and phosphorus in sewage can be diverted directly for reuse as a beneficial fertiliser.

Pros and Cons compared to other urine separation options

Pros:
- Can be used as part of a standard flush toilet infrastructure.
- Dual flush toilet uses less water than standard flush toilet.
- 'User-friendly' urine separation method.

Cons:
- Most suitable for new-build projects due to requirements for additional pipework.
- Still uses flush water, although greatly reduced compared to a standard flush toilet.
- Considerably more costly than standard flush toilets to purchase (offset by water charge savings).
- User education is needed to maximise the separation of the urine.

Urinals

Urinals may be used as part of a domestic sanitation system where urine diversion is desired, where budget may not permit a urine diversion dual-flush toilet and where the occupants are predominantly male. In this case, urinals can provide an easy way to recoup the fertiliser value of urine in a cost effective and user-friendly manner.

Pros and Cons compared to other urine separation options

Pros:
- Can be used as part of a standard flush toilet infrastructure.
- 'User-friendly' urine separation method (for boys and men at any rate).
- Less costly than urine separation toilets.

Cons:
- Most suitable for new-build projects due to requirements for additional pipework.

- Still use flush water, although greatly reduced upon a standard flush toilet. However waterless urinals are also available.
- Designs cater predominantly for men only. Although women's urinals have been designed and are in use, they are by no means as common as men's urinals.

Urine Diversion With Dry Toilets

Urine diversion is sometimes used in dry toilet designs as a way of keeping the faecal solids as dry as possible and low in volume. Typically the designs involve a urine-diverting seat that resembles the urine diversion dual-flush toilet design in that a front bowl collects urine for piping to an outside receptacle while the main container is for faecal solids.

Pros and Cons compared to other urine separation options

Pros:
- Solids are kept drier than with mixed dry toilet systems, so addition of fibrous material such as sawdust is not as crucial for the composting system to work effectively.
- For self-contained dry toilets, the volume of liquid runoff will be lower than where urine is included.

Cons:
- User education is needed to maximise the separation of the urine.
- Contamination with faecal solids is possible so the urine cannot be assumed to be bacteriologically safe without adequate maturation time.
- For thermophilic composting the nitrogenous element of the urine is important for generating the high temperatures for pathogen elimination.

Independent Urine Diversion

With the subject of urine diversion comes the realisation that long before any flush toilet infrastructure was invented there were hedges aplenty, and these were used with frequent abandon. Well, it may come as a surprise to many to realise that hedges are still prevalent today and just as useful as a way to convert urine into biomass as ever. Light hearted commentary aside, bear in mind that repeated urine application to one part of the garden is actively detrimental to the plants there and can also generate odours. Besides the trusty hedge, other 'independent' urinals include Mr Pee Buddy; the Straw Bale Urinal; and of course the compost heap, that nitrogen-hungry heap in the garden that simply thrives on attention.

Mr Pee Buddy is a DIY urinal developed by sailor Chris Melo to reduce flush volumes reaching the boat's holding tank, but can be used for any application where urine separation is required. The basic materials are easy to obtain and the design can be found in Carol Steinfeld's *Liquid Gold – The Lore and Logic of Using Urine to Grow Plants*.

The Straw Bale Urinal is exactly what it sounds like – a straw bale positioned on its side for use in a rain sheltered (and relatively private) place the garden as a urinal. The regular addition of nitrogen rich urine will break down the carbon rich straw and produce nutrient rich mulch that can be used in the garden or added to the compost heap. A straw bale will last about six months to a year with daily use.

Another very straightforward outdoor 'independent' urinal (also from Carol Steinfeld's storehouse of knowledge, *Liquid Gold*) is the compost heap. The regular daily application of urine will help to break down woody and fibrous compost material to help activate and heat the heap and produce good, nutrient rich compost for the garden.

Pros and Cons compared to other urine separation options

Pros:
- Direct use of fresh urine (from healthy individuals) is bacteriologically clean and is safe and easy to reuse appropriately on the garden. In this context, appropriately means either via a composting process with a carbonaceous material, or else used immediately on the garden at the base of growing plants, diluted approximately with eight parts water.
- Independent urine separation is the most cost effective way to divert urine away from being a source of pollution and towards being a source of fertiliser.

Cons:
- Direct use isn't to everyone's fancy – don't try this at work, folks! It doesn't go down too well in parks, car parks, or public areas either...
- Although independent urine reuse is an excellent way to keep urine separate from flush water, it isn't always suitable for everybody in the family, or for visiting guests. If you want everybody's contribution you may be better off with a urine diverting dual-flush toilet, for example.
- Independent urine separation is often an outdoor activity, so ease of use can be rainfall dependent, which can be a limitation to regular use.

Faecal Separator Systems

Faecal separators are used to separate faecal solids and toilet paper from the flush water in a standard sewer system. The most common commercial faecal separator for use with flush toilet systems is the Swedish Aquatron separator unit. The Aquatron unit separates the solids from liquids in flush water by a combination of gravity powered centrifugal force and surface tension. The separator unit diverts flush water out one side of the base of the unit while faeces and paper fall into the composting or storage chamber beneath the unit. Filtration methods have also been used, but are less common and are not included in the pros and cons listed below. While filtration methods have cost effectiveness and ease on their side, they are prone to clogging and need more space and maintenance than the Aquatron unit.

Pros and Cons compared to other separation options

Pros:
- Faecal separation can allow for composting of faecal solids within the framework of conventional flush toilet infrastructure. High on public acceptability from the perspective of the toilet user.
- May negate the requirement for septic tank emptying and associated costs and may even negate the need for a septic tank in certain circumstances. Such circumstances include sites where separation takes place below or adjacent to the dwelling, and where a constructed wetland is used after the unit prior to percolation to ensure that any solids that may get through the system during any malfunction are adequately retained in a suitable filtration system.

Cons:

- Although possible to retrofit into an existing dwelling, this system is best designed into a new-build project.
- A short pipe-run is preferable for this system, which works very well for Scandinavian houses with basements, where the Aquatron unit was developed, but not so efficiently for Irish houses which typically have septic tanks and other separation units further from the dwelling. The solution is to site the unit adjacent to the house in a sealed chamber.
- More regular maintenance is required in terms of emptying and cleaning than for a standard septic tank system.
- Compared to a standard dry toilet, the use of separation systems with standard sewer infrastructure still uses potable water as the transport medium, and requires subsequent treatment methods to clean up that water again before reintroducing it back into the environment.

Woodchip Filter Systems

This is a subset of Faecal Separator Systems, whereby solids are filtered out for composting using woodchips. Since the filter inlet feed may be pump macerated or otherwise quite broken up within the sewer line, this method may not be a source separation system in the same way as an Aquatron unit. However, it is closer to source separation than a septic tank because it is essentially a wet compost system (as described in Australian Codes if not Irish ones) which can be used for biomass and nutrient recovery.

Woodchip filters are generally contained within a purpose-built housing to allow separation to occur without additional washing by rainfall and to eliminate potential nuisance issues. Solviva designs in the US are available in Anna Eday's book Green Light at the End of the Tunnel. Eday recommends one enclosed filter (brownfilter) followed by a number of open planted woodchip basins (greenfilters) before a long planted woodchip filled percolation trench for final effluent distribution within the garden.

Also, vertical flow reed bed designs from French company Aquatiris include for a pump fed inlet without first using a septic tank. Thus the faecal solids are macerated and pump fed over the surface of the reed bed prior to additional filtration in a follow-on horizontal flow bed before discharge to ground via infiltration.

In both systems, compost worms play an active part in the humanure composting process while liquid is routed onward for further treatment and disposal.

Pros and Cons compared to faecal separation

Pros:

- Generally woodchip separators offer greater treatment than septic tanks because faecal nutrients are removed more effectively in wet composting systems than fully saturated environments (in which soluble nutrients are readily dissolved into the effluent).
- The organic matter accumulation and carbon sequestration is likely to be greater for wet compost and for septic tank sludge.
- Septic tank maintenance costs are not needed for these systems because all maintenance is possible to do manually when needed, without tankers.
- Humanure volumes shrink considerably within the system so maintenance is quite

infrequent; potentially up to 10 years between emptying intervals, depending on system size and usage.

Cons:
- As with the Aquatron, due to the head-loss within the filter system, pumping may be required where natural site topography is insufficient to facilitate gravity flows.
- Woodchip filter systems are not included in Irish or EU guidance to date (time of writing, June 2024), however section 1.3 of the EPA Code (2021) states that "... new and innovative products and technologies must be considered in detail by local authorities on a case-by-case basis..."
- As with the Aquatron, maintenance is non-conventional so needs to be carried out either by the homeowner, or by somebody who is willing to work with humanure.
- Some biomass is likely washed through the system. Although this will be removed in subsequent treatment stages, it means that the humanure compost from woodchip filters will be of a lower volume than from a compost toilet system.

5.7   Dry Toilet Systems

The lowest tech (potentially) and most straightforward source separation toilet is a dry system. Over the past fifty years many environmental education/research organisations have examined and experimented with alternatives to the flush toilet. The most recent initiative to push for a sustainable toilet was the Gates Foundation's "Reinvent the Toilet Challenge" in 2012. The principal reasons people usually have for seeking to avoid using flush toilets are typically summarised as follows:

- They are very wasteful of clean water resources;
- They are typically polluting of fresh water in the receiving environment;
- They waste nutrient rich organic matter and high-fertiliser value urine, which could otherwise be reintroduced into the soil.

As a result of this work there is a large research base and extensive wealth of experience throughout the world in the area of dry toilet systems. A quick Internet search will demonstrate the variety of suppliers on the market and the variety of different systems available.

Dry toilet systems are, as the name suggests, toilets that function without the use of water as a transport medium for disposal. Typically dry toilets are either compost toilets that contain and provide maturation time for a sizeable volume of humanure or faecal material, or are bucket systems that contain a comparatively small volume of material which is then removed to a remote composting location in the garden, or in some European countries, collected for municipal composting.

There are a number of different basic approaches used in dry toilet systems to achieve a safe, hygienic, manageable and environmentally sustainable alternative to standard sewered systems. The most basic dry toilet type (above and beyond no toilet at all, which is all too common around the world) is the outdoor pit toilet, the privy or pit latrine. This is not necessarily the most environmentally sound approach where site conditions are unfavourable, but can be suitable where soil depth and type are sufficient. A step up from that is a lined pit type toilet, where runoff is contained and contents are matured *in situ*. For an indoor alternative, there is the straightforward bucket system with a good thermophilic compost heap in the garden. At the most elaborate, there are fully self-contained composting units, which have electrically powered mixing/aeration bars that turn the organic matter within a plastic chamber so that the user empties finished compost for the garden.

Note that at the end of this section are a couple of dry toilet models that are mentioned which are not compost toilets at all. These include incineration toilets, which should carry a climate health warning for the amount of energy that they consume in the quest to avoid just opting for a sustainable dry loo. They are included here only because they are an option that exists on the market for import, should you care to go looking.

I haven't covered the area of outdoor flushless options since this book is geared specifically towards toilet solutions for household use. For outdoor adventurers straying a long way from home for extended periods, I refer the reader to Kathleen Meyer's *How to Shit in the Woods*.

Because the whole area of dry toilet technology is as varied as sewage treatment in general, I've included differentiating factors below to enable you to select the type of dry toilet that is most likely to meet your needs:

Indoor or Outdoor

Indoor systems have an obvious advantage that they are within easy access for regular use, particularly if they are the main toilet within the house. Outdoor systems have the advantage that they can be less costly to construct and require less frequent emptying. The potential for odour generation is less problematic in outdoor systems.

Self-contained Composting Types or Remote Composting Types

Self contained systems compost the faecal matter within the toilet itself, with the obvious advantage that when the time comes for emptying, the organic matter is already composted and more pleasant to remove; and the frequency of emptying is less. The advantage of the remote composting systems (or bucket systems) is that neither the toilet nor the composting location are necessarily fixed, and the space required for composting is delegated to the garden rather than taking up space in the house. Whichever type is selected, bear in mind that composting is an art in itself, and although very straightforward in many respects, does require attention to detail if you want to do it right. Joseph Jenkins' *Humanure Handbook* is an excellent reference book on this subject.

Electrically Powered Aeration, Manual Rotary Aeration or Stationary

Electrically powered mixers or fans have the advantage of minimising or removing the potential for odour nuisance and increasing composting speed and efficiency. They have the drawback of introducing an energy consumption element into the equation. Note that vents can also be added to some designs that work on solar convection rather than electric fans.

Electric Drying or Not

Some models work by drying out the faecal matter and urine completely, greatly reducing its volume and making emptying an easier and less frequent task. Such units tend to use a lot of electricity compared to most dry toilets, and this must be considered in the context of rising fuel costs and environmental impact. Incinerator toilets basically burn the contents after each use, and are only included here for completeness. They are not recommended as a sustainable sewage solution.

Urine Separation or Mixed

Since urine is basically bacteria free and very nutrient rich it makes a valuable liquid fertiliser, so keeping it separate from the faecal matter is an advantage for ease of use in farming or gardening. Generally the drier the compost in a dry toilet system, the easier it is to maintain, so separation of urine is typically recommended. However some self-contained composting units specifically require the liquid input from urine to create the appropriate composting environment.

Compost Chamber Size

Larger units generally need more strenuous maintenance at less frequent intervals than smaller storage units. However, with a good self-contained composting system, the longer maintenance interval does not necessarily mean a proportionally greater volume, since the composting process reduces the overall weight and volume of the organic matter. Space within the house is also a factor to consider when deciding upon system type in this context.

Maintenance Input

All the systems require emptying in some form or another. Self contained composting types require emptying of the compost; remote composting types require removal of faecal matter to a good outdoor composting system and then emptying of finished compost from there. Urine separation tanks require emptying when full. The urine tank capacity and frequency of use will clearly dictate the frequency of emptying and volume of each empty. Remote urine separator toilets such as the Nonolet require emptying every couple of days if in constant full family use, whereas piped urine separation tanks can have capacities sufficient for an entire development, emptied at the required interval by farmers for the fertilizer value of the liquid. Each system type has a different set of maintenance requirements and these must be carefully considered in the selection of a dry toilet system.

Fully Dry Toilet or Hybrid System

Some of the advantages of dry toilet technology can be used with a flush toilet system where appropriate. Aquatron separators, for example, separate flush water from solids to allow remote composting of faecal matter while at the same time enjoying the advantages of flush toilets within the house such as ease and familiarity of use. Urine separator toilets divert urine to an external urine storage tank while flushing away faecal matter and paper to an Aquatron, septic tank or sewer. Thus the urine can be reused on the garden or in agriculture, keeping about 50% of the nutrients out of the sewer system relatively easily.

Cost

There is a large variation in costs, varying from inexpensive DIY out-door pit systems to the most elaborate systems with associated price tags. Each person has a different budget and different set of priorities. Note that more cost does not necessarily mean a better system.

Bear in mind that these categories are relatively general groupings and that hybrids and variations often exist. Pick and choose the overall approach that looks most suitable for you.

A further consideration is the availability of public facilities for maintenance/composting. For example in Holland dry toilets are sometimes used on the canal houseboats. In one Dutch-designed example a urine separating, remote-composting toilet model (Nonolet Recreatie) is used and the resulting dry material is simply lifted out in a purpose made bag and placed in the local authority 'greens' bins for removal and municipal composting. This is only appropriate where it is acceptable to the local authority, and where the composting setup is sufficiently controlled for adequate breakdown of pathogens.

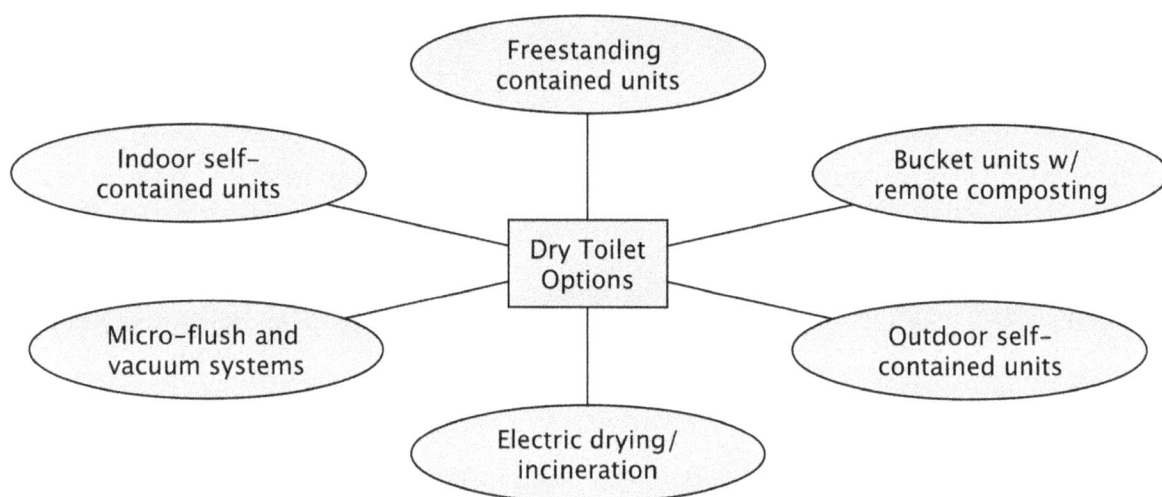

```
                    ┌─────────────────┐
                    │  Freestanding   │
                    │ contained units │
                    └────────┬────────┘
  ┌───────────────┐          │          ┌───────────────┐
  │  Indoor self- │          │          │  Bucket units w/ │
  │ contained units│─────────┤          │ remote composting│
  └───────────────┘     ┌────┴─────┐    └───────────────┘
                        │ Dry Toilet│
                        │  Options  │
  ┌───────────────┐     └────┬─────┘    ┌───────────────┐
  │ Micro-flush and│─────────┤          │  Outdoor self- │
  │ vacuum systems │          │          │ contained units│
  └───────────────┘          │          └───────────────┘
                    ┌────────┴────────┐
                    │ Electric drying/ │
                    │   incineration   │
                    └─────────────────┘
```

System layout within the site

Different types of dry toilet systems have different requirements in the context of the overall site layout. One fundamental difference between a sewered system and a dry toilet system is that there is no black water present. This means that a septic tank isn't strictly necessary. That said the grey water from the house still needs settlement and filtration prior to discharge to ground via percolation or to surface water via direct discharge. Although grey water still has pollution potential, it has the notable advantages of being essentially pathogen free, in sewage terms at least, and also lower in volume than standard domestic effluent. Grey water options are dealt with in the next section.

Pros and Cons compared to other treatment options

Pros:
- Dry toilets are much less wasteful of water than flush toilets.
- They have a much lower potential to pollute than standard flush toilet systems.
- They permit the recycling of human nutrients and biomass back to the soil, either in the garden or in agriculture.
- They can be much more cost effective to install and to maintain than any treatment system for flush toilets.
- Since they are dry systems, they can be used on sites of zero percolation and no discharge options. Grey water still requires attention, but is less of a challenge than a conventional sewage discharge.

Cons:

- Some people do not like the idea of using dry toilet systems. This psychological barrier is one of the biggest blocks to widespread acceptance of this eminently sustainable technology.
- Most dry toilet options have the potential for some odour generation.
- Most types require more user care, attention and maintenance than flush toilets to operate successfully. (If we ignore the need for septic tank maintenance...)
- Outdoor compost heaps have the potential to attract rodents if not carefully managed.
- Some types of dry toilet system have the potential for pathogen transfer to soil if not appropriately managed or if maturation times are insufficient.

Maintenance requirements for dry toilet systems

Different dry toilet systems require different degrees of maintenance, however the primary consideration is that every dry system requires emptying at some stage. Generally outdoor self-contained units require emptying every year or two, while bucket systems with remote composting require emptying every few days to a fortnight. The maintenance issues are listed separately for each dry toilet category to clarify the input required.

Indoor Contained Chamber Systems

As the name suggests, these are essentially pit toilets that are designed to be used as indoor toilets, hence the use of a contained chamber beneath the toilet stool. These toilets are designed to compost the faecal matter *in situ*. Emptying is typically only carried out every year or more.

Commercial models typically comprise of a large chamber located beneath the floor, in which the faecal material moves back, either by gravity or by mechanical means depending on the toilet design, towards an access door located outside the building. Examples include the Clivus Multrum and the Ecosan models.

Self-build indoor toilet designs often comprise two chambers built beneath the floor, and two toilet seats installed above them. One toilet seat has a cover screwed in place. The other toilet is in active use. After the first chamber has been filled, then the fixed cover is moved across to the other seat. The compost beneath is allowed to mature for the length of time it takes for the second chamber to fill. This is usually about a year, and depends on the chamber size and frequency of use. Access from outside the building permits emptying when necessary.

Some commercially available chamber systems use one composting chamber for several toilet seats or even for whole toilet blocks. Some public project dry toilet applications work better than others, so it is very important that the users understand the limitations, and that adequate maintenance and inspection are used to keep everything working well.

Pros and Cons compared to other dry toilet options

Pros:

- Having self-contained chambers, these toilets require less frequent maintenance – and the final material tends to be finished compost rather than faecal matter and so is lighter and easier to work with.

- Indoor units generally present the obvious advantage of being more accessible and easy to use day to day.

Cons:
- The cost of large chamber units tends to be higher than small indoor bucket models. Bear in mind that indoor chamber units can also be self-build, which can offer distinct cost savings.
- Some commercial chamber units recommend a fan to contain odours, which can mean that electricity consumption, however small it may be, may be necessary. Some indoor ventilated units, however, work using sunlight to heat a vent pipe to provide a downward draw through the toilet seat. Also, solar panels are available from some suppliers to power electric fans.

Freestanding Self-contained Composting Units

Some models comprise a large box within which baffles and filter grates provide for aeration of the faecal material as it moves slowly towards a bottom collecting tray. This unit sits directly on the floor without any particular plumbing requirements other than the installation of a vertical vent pipe through the wall or roof.

Pros and Cons compared to other dry toilet options

Pros:
- Relatively easy to set up, without requirements for digging or plumbing other than drilling for the fan vent pipe.
- Some or all composting takes place within the toilet unit, making emptying less frequent than bucket systems.

Cons:
- Relatively expensive to purchase compared to bucket type toilets with remote composting, without necessarily offering corresponding benefit over other models.
- Doesn't necessarily provide full composting if used on a full-time basis, and if overloaded, can just fill up and require a full clean down inside.
- Emptying can sometimes be awkward to access and carry out, depending on the model selected.

Bucket Units With Remote Composting

Remote composting toilets are basically bucket toilets that hold faecal material, paper and urine for regular emptying to a separate location for composting. There are many different types available to purchase and different designs that can be followed if you want to make your own.

Some toilets specifically separate urine in order to minimise odour generation and avoid the use of an additive such as sawdust for soakage. Other designs, such as Jenkins (2005), specifically recommend including urine in the final compost to ensure good breakdown of carbonaceous material in the compost heap.

By way of examples, there are the Dutch Nonolet toilet, some of the Swedish Biolet and Separett models, and Joseph Jenkins' "lovable loo". The Nonolet, designed by the environmental charity *De Twaalf Ambachten*, uses urine separation to keep the solids as dry as possible. After use, contents are compacted slightly with a paper towel and flat plunger-like implement to make most efficient use of space and to dry out the contents further. These can be placed in a sealed unit such as a dedicated bin for worm composting, or in a garden compost heap following the general principles of *The Humanure Handbook*.

Just as a flush toilet requires the backup infrastructure of a septic tank and percolation area as a minimum, so bucket units with remote composting require an outside location for both composting and maturation. This compost area can be an enclosed plastic unit to ensure that no mixing with the environment occurs prior to maturation, or can be a straightforward thermophilic compost heap. *The Humanure Handbook* is the essential reference book for anybody using this type of toilet, with clear instructions and common sense, practical advice and insights.

Pros and Cons compared to other dry toilet options

Pros:
- The primary advantage of remote composting systems is that they can fit into almost any space in the home and function effectively without any plumbing requirements. Most models do not even require vent pipefitting.
- They are very cost effective – probably the lowest cost toilet system available, and can be home-built for almost zero cost or purchased ready made for less than the price of a septic tank.

Cons:
- Remote composting toilets require on-going regular maintenance, emptying the toilet bucket to the compost area and then occasional maintenance of the compost area itself.
- They don't offer the same invisibility as a deep pit compost chamber, so you can't ignore the straightforward reality of the contents. However most units require the addition of some carbonaceous cover material after each use, which contains odours and, I would argue, replaces the psychological satisfaction of flushing afterwards.
- Indoor bucket type compost toilets do smell slightly, although they can be fitted with a convection vent if required and some commercial models come with a vent or charcoal filter and fan to deal with odours.

Outdoor Compost Toilets

Outdoor compost toilets are the category of compost toilet that many people think of in the context of 'compost toilets' generally. These are the pit latrines and the outdoor privies; the toilets that were commonplace in Ireland only a few generations ago. Many people still remember them from home or school, without particularly fond memories. There are of course, many modern versions now available on the market that work very effectively, producing good quality compost in a safe, clean, sanitary way.

Typically outdoor toilets are sited in an independent outbuilding and require emptying only infrequently. For some outdoor toilets, the building is simply moved after a number of years and a tree can be planted over the contents after backfilling with soil. Look up Tree Bog online for photos of a straightforward outdoor toilet that relies upon hungry willows and other trees planted around it to absorb the abundance of nutrients and protect the groundwater beneath.

Not all sites are suitable for an unlined outdoor toilet, so make sure your soil is suitable before you construct, or use a system with a contained base. For an outdoor toilet with a contained base, see John Seymour's Thunderbox toilet described in *The New Complete Book of Self-Sufficiency* (2003). This system can be built into the house at the initial construction stage (fitting it into the category of an indoor contained chamber system) if required, but is difficult to retrofit into an existing house. However it is classified here as an outdoor type since that is its easiest and most common application.

For a humorous account of the challenges and ingenious responses of a 'champion privybuilder', *The Specialist* by Charles Sale (1929) raises valuable lessons for the renaissance of this traditional toilet model.

Pros and Cons compared to other dry toilet options

Pros:
- Outdoor systems have the advantage of being somewhat more forgivable for any potential odour arising. Being outdoors, they can easily be sited in well ventilated structure to negate, or minimise odour nuisance. That said, many designs overcome odours very effectively.
- As with the large indoor contained chambers, contained outdoor types require less frequent maintenance intervals, and the material removed is more like compost than faecal matter.
- Outdoor units can be easier to empty than indoor units since it is somewhat more acceptable for compost to spill occasionally onto a lawn than a bathroom floor. This advantage also applies to indoor units that have outdoor access for emptying.

Cons:
- Outdoor toilets have the obvious limitation of being less accessible for easy use, particularly during wet weather and at night time.
- Unless it is the only toilet serving the house, an outdoor loo typically won't receive as much use as an indoor equivalent, particularly at night time. This can reduce the overall beneficial environmental value of selecting a dry toilet, and can certainly reduce the amount of urine and compost that will be generated.
- Maintenance, when it is needed, is a relatively labour intensive activity requiring physical strength and attention to clean handling techniques.
- Unlined toilets can cause groundwater pollution in the same way as a poorly laid out septic tank and percolation area if constructed in inappropriate conditions.

Micro-flush and Vacuum Flush Systems

Micro-flush and vacuum flush toilets use a small volume of water to deliver the contents of the toilet to a remote composting chamber. The flush water is typically about 0.5 litres, compared with 3 or 6 litres per flush for a dual flush water saving toilet or 9-12 litres for a standard flush toilet. This small volume of water is typically drained off from the bottom of the compost chamber for treatment and disposal, allowing the dry compost process to continue in the chamber.

Pros and Cons compared to other dry toilet options

Pros:
- For people who want a flush toilet, but also want to minimise water use and compost

humanure, micro-flush and vacuum flush toilets can allow a 'nearly dry' flush toilet option.

- This option can work for sites where a flush toilet is wanted, but where site conditions and environmental regulations make discharge to the environment problematic.

Cons:
- Electricity is needed to run the system.
- Costs are higher than most dry toilet systems due to the different components involved: chiefly the toilet unit and the composting unit.
- The residual water volumes still need treatment.
- These systems are not common in Ireland, so visiting working examples and talking to satisfied users is difficult.

Electric Drying and Incinerator Toilets

Electric drying toilets use a fan and heater combination to dry the urine and faecal matter so that the resulting contents of the dry toilet chamber are lighter and reduced in overall volume. This means that less frequent emptying is necessary. Like all remote composting systems, these toilets require emptying to a remote composting location, or more commonly in the case of electric drying systems: burial or disposal.

Incinerator toilets go a step further and literally burn all of the contents. Emptying is clearly less frequent, but carbon footprint and cost rise accordingly. This option may have the environmental advantage of eliminating water pollution from toilets, but the increase in energy usage cannot really be said to be sustainable. Annual running costs at the current cost of fuel (2014 prices) would run to about €2000/yr plus maintenance and the replacement stock of bowl liner sheets that are needed with every use. That makes for a fairly expensive option when all you really wanted was to spend a penny...

Pros and Cons compared to other dry toilet options
Pros:
- Electric drying toilets require less frequent emptying than other remote composting toilets.
- They may be used on an infrequent basis without odour generation.
- Incinerator toilets reduce the total volume to ash, for ease of emptying and disposal.

Cons:
- Electric drying toilets use more electricity than standard compost toilets due to the need to dry out all urine and faeces.
- Incinerator toilets working on electricity use about 1kWhr per use. So, if I left a 1000W hair drier on for an hour, that would be one use, then went to the loo again a couple of hours later... and so on. That's quite a lot of electricity. For propane or diesel powered incinerator toilets the fuel usage for a family of four would be about 2500 litres/yr. I'm not sure that these quite qualify for the eco toilet of the year award, but they are here as an option for consideration nonetheless – if you want to up your fuel usage in an age of climate change and energy uncertainty and pay more in annual running costs than any other toilet listed in this book. Actually, please don't. I'm not sure I'd like to have any part in it, and having written the book, I'd be implicated automatically.
- Since most dry toilets have the advantage of providing humanure for your garden at the end of the year, it is fair to say that while the electric drying toilets could

be rehydrated for composting, there is not much humic value in incinerator ash – whatever its origins.

5.8 Grey Water Options

'Grey water' is the term used to describe all the water from sinks, wash hand basins, washing machines and dish washers, showers and baths. Basically everything in the sewer pipes except the black water from the toilets. Storm water, from roof surfaces, as well as roads, driveways and footpaths, should be kept separate from the sewers and disposed of directly to percolation (not the same percolation area as your septic tank effluent!) or via filtration through a suitable SuDS (Sustainable Drainage System) unit, such as a constructed wetland, prior to discharge to ground or surface water. Grey water comprises about half the volume of domestic sewage, and due to the presence of cleaning chemicals, detergents, cosmetics etc. it accounts for almost all the toxins in sewage. Grey water can also have high phosphate concentrations where these are present in washing powders.

Under the current EPA *Code of Practice*, it is recommended that grey water should accompany black water to the septic tank or wastewater treatment system. While this has clear advantages in that it sets out to eliminate both grey water soak pits and connecting grey water pipes to storm drains from roof surfaces – both of which can cause pollution in the receiving environment – it does mean that grey water recycling and reuse are typically overlooked as viable treatment/management options. Grey water recycling not only has the potential to save on water costs, but also provides a way to minimise the volume of sewage for disposal. This is particularly useful where a large percolation area is needed for heavy soils.

Summer water shortages in Dublin have prompted Dublin City Council to put grey water recycling guidelines on their website to help homeowners to minimise their overall water consumption. Such recommendations are common in the US and Australia, where water shortages are part of daily life. Whether we have direct water charges or taxes, we still pay for our water and municipal sewage treatment. The more we conserve water the less the council has to spend filtering drinking water, adding chloride and fluoride, pumping it to our homes, and then in sewered areas, pumping it away again and treating it before discharging it to a river or the sea...

Other than civic duty *vis a vis* the possibility of reducing costs to the council, why would I want to deal with grey water separately from black water? The most common reason for wanting to separate grey water from sewage is a desire to conserve water for environmental reasons. The next most common reason in my experience is if there is no black water, i.e. where a dry toilet is used instead. Another reason is where an old septic tank is working effectively, without grey water entering it, and the homeowner wishes to preserve the effective microbial environment in the tank by keeping out the grey water, but is under pressure to combine the two for *Code of Practice* reasons.

There are a number of approaches that can be adopted for grey water disposal, as follows:

- Combining with black water to your sewage treatment system (dealt with as part of the options outlined in other sections in this book).
- Recycling grey water for toilets and washing machines after filtration.

- Reusing grey water for garden irrigation, with or without filtration.
- Discharging to ground via percolation, independently of your septic tank system.

One way or the other the grey water will need some degree of filtration or separation before recycling or discharge. For whatever portion doesn't go to your septic tank, grey water treatment options include the following:

- A proprietary grey water treatment system for recycling.
- An equivalent purpose-built filter or even rudimentary screen prior to reuse depending on the application.
- A constructed wetland or reed bed.
- Soil percolation.

There are two important considerations with any grey water recycling system. Firstly, when grey water remains in a storage tank for more than a few hours, it will typically begin to become anoxic. The anaerobic conditions have the potential to create odours and discolouration, which reduce the recyclability of the grey water. Direct feed recycling, such as diverting wash hand basin outlets into the toilet cistern, can avoid this challenge.

Secondly if you reuse your grey water as irrigation water on your garden, the ingredients in every single product you use in your home suddenly become part of the food chain. This is particularly important for food plants, whether they be root vegetables or walnut trees. If you use cosmetics, cleaners or detergents with noxious chemicals in them, these will be taken up to some extent in the plants, and a portion of them will end up back in your diet. Paint, oils, chemicals and the like from hobbies or work may also find their way into your grey water system through hand washing and washing down equipment. Due care should thus be taken with any grey water irrigation system to protect against inadvertent or careless contamination of your food supply. The easiest way to overcome this challenge is to allow the dirtiest grey water to go with the black water, and divert the cleanest water for irrigation. Thus, washing machines and dish washers can be routed to the septic tank while baths and showers can become a source of water for the garden, taking care to only use health food shop shampoos and conditioners.

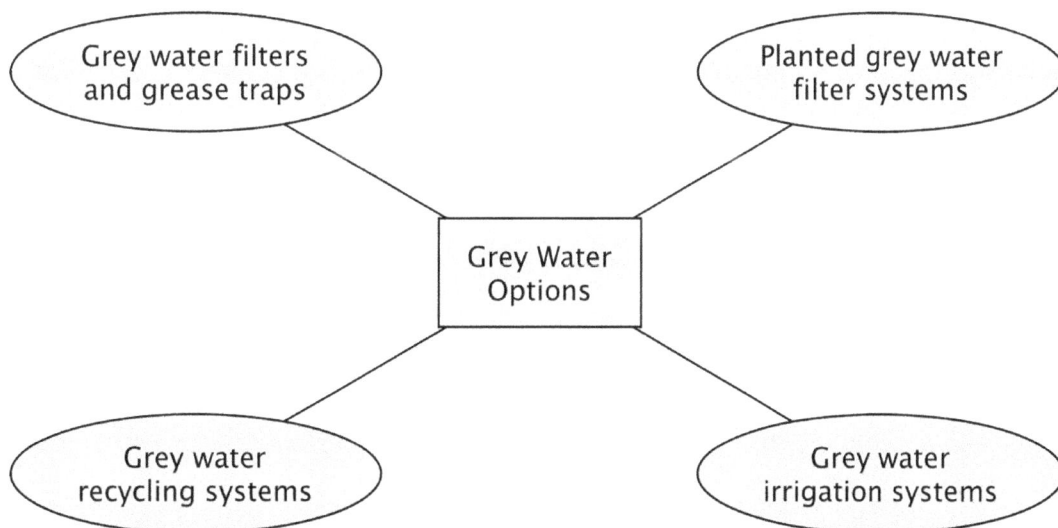

System layout within the site

Grey water treatment and disposal options are many and varied, ranging from permaculture techniques to proprietary recycling units, which filter grey water for reuse in toilets and washing machines. As such it is just as difficult to offer a concise description of their layout within a site. However by way of a summary description we can take the different types, as follows:

- Where grey water is combined with black water, there is no additional infrastructure required.
- Where grey water is recycled for toilet flushing and washing machine use there is plumbing infrastructure required within the house, and a filter system to keep the water fresh enough for reuse, but there are no additional external site works required.
- Where grey water is reused in the garden the infrastructure can vary from a simple hosepipe feeding from the bath to the polytunnel, to a reed bed filter system leading to a storage pond.
- Where grey water is discharged to ground, this can be done by means of a conventional septic tank set-up and percolation, or by means of a dispersal system using woodchip percolation pockets which combine disposal and irrigation.

The best way to determine the site set up requirements is to select a system and see if that general approach fits within the overall site constraints. Grey water treatment/ disposal is typically less difficult to fit within a site boundary than sewage. Note that not all practical grey water management techniques are specifically permitted under Irish government guidelines. In fact, the only route that is recommended in the *Code of Practice* is combined flow with black water. As such, it is important to approach the topic carefully to ensure legal compliance and also to make sure you don't inadvertently cause groundwater pollution.

Pros and Cons compared to piping grey water directly to the septic tank

Pros:
- By keeping the grey water element of your wastewater separate from black water you maintain the possibility for reuse and recycling of this portion of the water from the house, which can then be used for toilet flushing or irrigation to maximise water conservation.
- Grey water typically contains many of the heavy metals and other toxins of raw sewage, and so by keeping it out of the septic tank you can help to keep the tank microbial flora healthier and more effective.
- Separate grey water filtration and disposal becomes a necessity rather than an advantage *per se* if the selected toilet system is a dry compost toilet. The advantage of a grey water system in this context rather than a standard septic tank and percolation system is that the infrastructure required for grey water and for black water is slightly different, and cost and carbon savings may be made by avoiding the septic tank and extensive piping requirements of a standard system.

Cons:
- The current legal frameworks point builders directly towards combining grey and black water, so the methods of separating these elements of domestic wastewater are not extensively investigated in Ireland and a general lack of solid knowledge about separate grey water management can lead to both mistakes in application

and/or more cost than necessary.

- If a septic tank system or treatment system are already in use or selected for use, then a separate grey water system will be an additional expense and land take that may not otherwise be needed.
- Reuse of grey water on food crops requires careful on-going vigilance in that your shopping basket needs to contain only products that will be safe for putting on your veggies. A desire for an environmentally friendly sewage treatment usually coincides with a desire to have a healthy lifestyle, so this drawback is not usually a problem, but more of a reminder.

Maintenance requirements for grey water systems

Grey water systems require maintenance like any other treatment component. The principle issue is as per black water systems, namely suitable solids separation prior to treatment and disposal.

- Empty the grease trap (if used) on a regular basis to avoid oil/fats/greases and settled sediments from carrying over to the treatment section of the system used.
- Inspect the system at a frequency suited to the option selected (typically monthly to annually depending on system type) to check for blockages, overflows or malfunctions.
- In the case of planted filters, check plant health on a quarterly basis.
- Some planted systems require harvesting and removal of the plants to maintain continued nutrient uptake and removal from the treatment area (and to offer the benefits of nutrient recycling).

Package Filtration of Grey Water for Non-potable Domestic Recycling

One option for grey water is to use a proprietary filter system, which is designed to remove bacteria and suspended solids so that the water can be reused for toilet flushing, washing machines and/or garden irrigation. If you want to pursue this route, take care that the system you buy actually does what you want it to do. There's no point getting a system designed to produce an effluent that is suitable for watering your flowers with if what you want is a high degree of reuse in toilet cisterns and washing machines. An Internet search for something like "grey water recycling filters" will show a good spread of the systems available; but narrow it down to UK searches only if you don't want to fly your system across the Atlantic.

Pros and Cons compared to other grey water options

Pros:
- This approach can save considerable amounts of water, particularly if used for all domestic non-potable applications, such as toilet flushing, washing machines, dishwashers and outdoor wash water.
- If you are using a zero discharge willow facility, constructed wetland or reed bed for your septic tank effluent treatment, the reduced grey water volumes will either allow a smaller treatment system to be used from day one, or will allow the wetland system to achieve a higher water quality.

Cons:
- The cost of setting up and maintaining a package filter system for grey water is not

inconsiderable. The cost for full recycling of all bath and shower water for reuse in toilets, washing machines and outdoor taps can be several thousand euros or more.

- Electricity is needed to power the pumps for the filter system and/or the recirculation pump.
- A separate plumbing infrastructure is needed to accommodate the system, so it is best done as part of a new-build project. On this note, if you have a very large house, it may be more environmentally friendly to use just the one set of piping and treat grey and black together rather than buying twice the amount of PVC piping.

Grey Water Diversion for Irrigation and/or Immediate Reuse

Another grey water option is to send the cleanest grey water – from baths and showers – to an irrigation system. This may be used directly on the garden or polytunnel, or may be filtered and stored for use within a day or two. Filtration in a reed bed, constructed wetland or other filter is necessary to keep stored grey water relatively fresh.

Diversion is a flexible term. At its simplest it can be reusing fairly clean wash-up water on your roses or running off the cold water into a jug for the garden (or house plants) before the hot tap warms up enough. A diverter can be bought which fits onto grey water outlet pipes so that at the pull of a string you can decide whether your bath/shower water should go to the septic tank or the garden. A simple reuse design involves routing the wash hand basin outlet back to the toilet cistern so that after using the loo, you wash your hands and begin to refill the cistern again. One company, Caroma, has taken this idea to its logical conclusion and made the basin into the lid of the toilet cistern.

If you want an inexpensive filter system for grey water, a sand filter drum can provide good filtration prior to irrigation. Look online for ideas and designs, or contact an environmental consultant or permaculture designer for guidance.

Pros and Cons compared to other grey water options

Pros:
- This is the easiest, lowest cost grey water recycling option.
- A significant volume of your grey water can be diverted almost directly to reuse, leaving your filter system or main treatment system to cater for a smaller volume.

Cons:
- Your garden may not need all of the water generated, and roof water can often fulfil most garden needs.
- It can be easier to lump all of your grey water into the one treatment and disposal method rather than having lots of different systems in your garden.

Constructed Wetlands and Reed Beds

Constructed wetlands and reed beds can be designed to filter the grey water so that it can be reused in the garden for irrigation, for diversion to a garden pond or disposed of via percolation. The size of the wetland is a function of how clean or dirty the grey water is coming in and what degree of treatment is needed. For routing to a pond, quite a large wetland (c.4m x 12m) may be needed in order to ensure that the water is sufficiently clean to avoid regular cloudiness and algal growth. This is smaller than a standard sized wetland for dealing with grey and black water combined, but is larger than many books may suggest. That said if you are filtering relatively clean grey water for reuse around shrubs etc. then a couple of m^2 (if even) may well suffice. It really depends on the application.

Pros and Cons compared to other grey water options

Pros:
- The cost can be lower than a package filter unit.
- There is no electricity needed unless you are working against gravity either for treatment or for reuse.
- They can provide an attractive habitat for wildlife within the garden.

Cons:
- The space requirements can be more than some small gardens can spare.
- Any filter system is more costly than simply routing grey water to percolation.
- Constructed wetlands are open systems, with potential for some odour generation and potentially open water as a safety hazard. Gravel reed beds can avoid these issues, but have the drawback of requiring quarried gravel and also an increased potential for blocking with sediments.

Grey Water Percolation Systems

If you are using a dry toilet option, or do not currently wish to route grey water to your existing septic tank, then possibly the most straightforward way to dispose of grey water is to filter it in a percolation system. The EPA *Code of Practice* figures of hydraulic loading rates can be applied, taking approximately 60% of the total volume. Thus by way of example, if there are 5 people in the house the grey water percolation area requirements will vary from ≥22.5m^2 for free draining soils to ≥150m^2 for poorer draining soils. This assumes that the soil percolation rate is sufficient, and that all other soil characteristics, such as depth, are suitable. For a trench system this works out at a range between ≥18m and ≥57m of linear percolation trench length. These figures are for treated effluent – so grey water after filtration through a reed bed or wetland. For direct discharge to percolation after settlement in a grease trap or septic tank allow about ≥54m of piped trench length for soils with a t or p value of up to 50 minutes. Standard percolation sewer pipes are designed for relatively clean effluent and grey water needs filtration prior to disposal in this way; otherwise the pipes may get clogged with sediments and cease to function. Even settlement in a grease trap will help to minimise sediment accumulation in percolation pipes, so full filtration in a reed bed or wetland isn't strictly necessary if soil conditions are suitable.

Note that percolation areas for grey water are not the same as soak pits. Soak pits involve the discharge of grey water over a very small surface area at considerable depth. These factors limit the natural capacity of unsaturated soil to filter the water before it reaches the groundwater or bedrock. The use of soak pits to dispose of untreated grey water is not acceptable environmentally or legally, and the *Code of Practice* has basically stopped this disposal method from being continued as a common practice.

Pros and Cons compared to other grey water options

Pros:
- Percolation systems are a known quantity for builders, planners and engineers, so it

may be easier to get planning permission for this type of grey water treatment than other approaches. That said, if you are applying for a separate grey water system then you are already outside the box, so you may as well carefully examine your priorities and propose what you actually want rather than what you think may be acceptable.

- Properly constructed and maintained, a grey water percolation area can give years of trouble-free treatment.
- Percolation is an out of sight approach – buried in the ground and out of mind (for better or for worse).

Cons:

- The *Code of Practice* sizing is a lot larger than a traditional grey water soak pit, and given the nature of much of what goes down the drain this is entirely justified. However if you are keen on both water conservation and careful with what you dispose of, it is possible that the *Code* is more generous on land area than needed. Then again, if this applies to you, chances are that percolation is also considered simply a waste of precious water...

Planted Grey Water Percolation Systems

Planted percolation systems are set out in approximately the same fashion as ordinary percolation systems except that the area is planted with a crop such as willow trees for coppicing or comfrey for harvesting as a compost ingredient. The distribution system needs to be able to accommodate active root growth without clogging, so a leach-field infiltration chamber layout may be more appropriate than standard sewer piping. Infiltration chambers are essentially large half-pipes which allow the water to spread over a flat percolation area and drain down into the soil. The advantage is that they don't have holes for invasive willow roots to block up.

Note that planting over percolation areas should only be carried out where suitable subsoil depths occur. In areas with thin subsoil over vulnerable aquifers, for example, root growth can lead to preferential flow paths within the soil, short-circuiting the treatment benefits offered by the soil and roots themselves.

Willow trees (*Salix viminalis*) can be harvested on a rotation basis every 3 years to provide small diameter logs for firewood or wood-chips for fuel or landscaping. Except for phosphate detergents, there is a relatively low nutrient content in grey water, so while the willows will help to keep your groundwater cleaner, the trees themselves will not perform as effectively as trees in a zero discharge willow facility for sewage treatment for example. Nonetheless, use a biomass hybrid that is bred for the task of maximising biomass accumulation and maximising water and pollutant uptake.

Comfrey (*Symphytum officinale x uplandicum*) can be planted through a purpose-designed grey water percolation area to recoup nitrates, phosphates and potassium through their deep roots. The leaves are then cut back between once and three times in a growing season and used as a compost ingredient or as a direct mulch layer around fruit trees. This method recycles nutrients in the grey water back to the garden in a way that eliminates any contact between the water and the crops. Bear in mind however that if you use cosmetics, detergents or household chemicals then some of the ingredients are likely to accumulate in the leaves and be recycled to your compost heap along with the beneficial nutrients.

Pros and Cons compared to other grey water options

Pros:
- Planted systems afford the possibility for nutrient capture and recirculation to the garden or generation of firewood, depending on the system chosen.
- The planted nature of the system also provides additional treatment, so the large EPA *Code of Practice* size stipulations may not need to apply in all cases.

Cons:
- Due to the relative novelty of these systems in Ireland, design sizing recommendations are best kept at EPA *Code of Practice* sizes, which means that the systems will be larger than may be necessary.
- This lack of clear sizing protocol means that purpose designed systems may be somewhat more costly than would otherwise be the case, due to consultancy and design inputs required for planning and construction.
- Preferential flow paths caused by root ingress have the potential to promote excessive movement through the soil, and may lead to groundwater contamination. However, the presence of a root mat may also provide better filtration of nutrients than would otherwise be the case.
- Maintenance is required. Willows need to be cut back every three years. Comfrey is typically given two or three cuts per season. If these are viewed as a resource, then this maintenance becomes a bonus rather than a chore.

5.9   Tertiary Treatment Systems

Tertiary treatment systems are generally used where enhanced environmental protection is needed, whether due to site characteristics or receiving environment vulnerability. In the first instance, if the soil percolation is too rapid, then a tertiary treatment system is a necessary component of the overall treatment process, to ensure that the groundwater is adequately protected. In the second instance, if the site characteristics are acceptable, but the receiving environment – be it a vulnerable aquifer or an important river habitat – requires additional protection from effluent discharges, then a tertiary treatment system can be used to enhance the final effluent quality.

Tertiary treatment typically provides enhanced removal of BOD and suspended solids, as well as reduction in nitrogen and phosphorus compounds in the effluent. Bacterial numbers are also typically reduced. Some tertiary treatment systems are specifically designed to remove a particular pollutant, such as phosphorus removal by dosing with ferric sulphate, or bacterial sterilisation by UV light.

Initial description

Treatment systems described in the EPA *Code of Practice* include the following:

- Soil polishing filters
- Sand polishing filters
- Treatment wetlands
- Packaged tertiary treatment systems of various descriptions

```
                                                    ┌──────────────┐
  ( Soil polishing )                                ( Sand polishing )
  (    filters    )                                 (    filters     )

                      ┌───────────────────┐
                      │ Tertiary Treatment│
                      │     Systems       │
                      └───────────────────┘

  (  Treatment  )                                   (  Packaged  )
  (  wetlands   )                                   (  systems   )
```

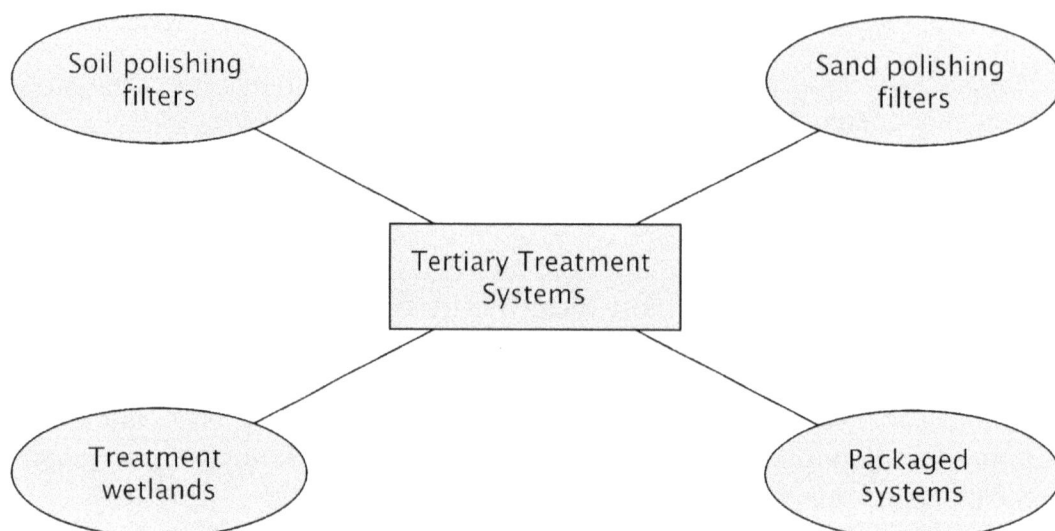

System layout within the site

Tertiary treatment systems are generally used to polish secondary treated effluent. In other words, the typical layout would be a secondary treatment system such as a package treatment system or reed bed, followed by a tertiary treatment system, followed by the selected disposal option for the final effluent. For soil and sand polishing filters the disposal route is typically, although not always, coupled with disposal to the groundwater beneath it.

Where a packaged tertiary treatment unit or treatment wetland is used, the disposal route can be directly to a distribution area rather than a larger soil polishing filter or percolation area. Although distribution area, polishing filter and percolation area are all essentially percolation systems, their difference is the size and layout, dictated by site conditions and effluent quality requirements.

While a standard septic tank and percolation area is acceptable in the *Code of Practice* in soils with t values between 3 and 50 minutes with unsaturated ground conditions of 1.5m depth beneath the pipe invert (1.2m below the trench base), secondary treatment systems widen the range of soil conditions to t≤90 minutes and a p value of between 3 and 75 minutes. The depth of soil can also be reduced to 0.9m. Tertiary treatment systems may then be added to reduce the soil depths to 300mm over bedrock or groundwater.

Pros and Cons compared to a percolation system without secondary and tertiary treatment

Pros:
- Tertiary treatment systems improve the overall water quality prior to discharge and thus help to protect the receiving surface water or groundwater.
- Tertiary treatment systems broaden the range of soil characteristics that are acceptable for discharge, enabling sites of poorer characteristics to be considered acceptable. See table 6.3 of the EPA *Code of Practice* for a more detailed interpretation of percolation test results; reproduced in Appendix 6. See also table 10.4, addendum to the EPA *Code.*

Cons:

- Tertiary treatment systems are an added expense where soils are otherwise suitable.
- Most tertiary treatment systems also take up additional space within the garden layout. Exceptions to this are direct discharge soil percolation systems that underlie an intermittently pumped secondary treatment system and some combined package secondary/tertiary treatment systems.

Maintenance requirements for tertiary treatment systems

The principle maintenance requirements of any sewage treatment system are the regular desludging of the settlement system. If tertiary treatment systems of any sort get overloaded with oils/fats/greases or suspended solids, then their lifetime will be significantly shortened. Sludge loading to open systems, such as soil based constructed wetlands, are much easier to remedy than enclosed systems such as soil polishing filters. In short, all maintenance requirements for septic tanks or secondary treatment systems preceding the tertiary treatment component should be strictly followed to ensure that the tertiary system functions effectively for its design lifetime.

Where the tertiary treatment system requires addition of chemicals such as ferric or aluminium sulphate for phosphate removal or chlorine for sterilisation or the use of UV bulbs or electrical inputs for ozone sterilisation, then it is crucial to the proper functioning of the system that these chemicals or components are added as needed and working correctly.

Soil Polishing Filters

Soil polishing filters basically consist of ≥0.9m of unsaturated soil of suitable percolation characteristics, overlaid with a shallow gravel distribution layer, onto which the treated effluent is introduced. Three different configurations are outlined in the *Code of Practice*: namely *direct discharge* where effluent is introduced directly from an intermittent filter (for example a pump-fed coconut filter) positioned above it; *pumped discharge* where effluent is pumped to a raised mound; and *gravity discharge* where effluent flows through a series of perforated pipes in a similar manner to a standard percolation area.

Pros and Cons compared to other polishing filter options

Pros:

- These are the most commonly built polishing filters and thus are typically the most cost effective to install since builders do not usually need outside design input or product purchases other than gravel and pipe fittings.
- The gravity discharge option does not require a pumped feed, but is only suitable where falls permit.

Cons:

- One of the drawbacks of being the most commonly installed system is that while everybody thinks that they can install them, not everyone does it with the care and attention to detail that are required of any system to get it to work at maximum effectiveness.
- Both direct discharge and pumped discharge soil polishing filters require a pumped feed somewhere within the overall treatment process, and occasionally two pumps where needed. Gravity dosing systems also exist, to provide a pulse flow to the

tertiary polishing filter. While these remove the electricity cost, they are expensive to buy and typically need more frequent inspection and maintenance than a pump to keep them working effectively.

Sand Polishing Filters

Sand polishing filters are set up in much the same way as pumped soil polishing filters. The effluent is introduced over a constructed bed of sand of specific characteristics which then provides secondary treated effluent with additional polishing prior to discharge to surface water or to groundwater via a discharge zone.

Pros and Cons compared to other polishing filter options

Pros:
- Sand filters may receive effluent discharges of up to 60 litres/m²/d. This is three times higher than the highest dosing rate onto direct and pumped discharge soil polishing filters for free draining soils, and 20 times the rate for soils with slow percolation rates (of t between 51 and 75 minutes). This makes sand filters between 3 and 20 times more efficient in terms of space within a site.

Cons:
- Although sand filters are more space efficient than soil polishing filters, they also require a discharge zone for soils of p/t between 20 and 75 minutes. According to table 10.4 of the EPA *Code of Practice* (Clarification Document) the area requirement is between 12.5 and 47m² for a household of 5pe.
- Obtaining sand of suitable quality is crucial for the on-going effectiveness of the filter, and this may be difficult or expensive to obtain. Without the use of suitable sand the filter is likely to allow effluent through too quickly and be ineffective; or too slowly and clog up.
- Sand polishing filters require a pumped distribution feed.

Constructed Wetlands and Reed Beds

Constructed wetland and reed beds are described in more detail in the secondary treatment section of this book, but they can also be used for tertiary treatment of effluent. In this context they are used as add-on polishing filter systems following secondary treatment. A low-tech, high-spec system could well include a septic tank, a soil-based constructed wetland sized for secondary treatment (on indigenous marl clay subsoil, without a plastic liner) followed by another such unit for tertiary treatment, followed by a distribution area. Four different constructed wetland types are listed in the *Code of Practice*, as follows:

- Horizontal-flow gravel reed beds
- Vertical-flow gravel reed beds
- Vertical-flow sand reed beds
- Soil-based constructed wetlands

Vertical-flow reed beds require intermittent dosing of effluent across the surface of the reed bed to be effective. A pumped feed is the most common type of dosing method, but an alternative system such as a gravity-dosing box or bell syphon system may be used where falls allow. The head-loss through the dosing system and the vertical-flow reed bed can be approximately 1.5m. These components are more costly than a pump, but can then perform the same dosing function without the on-going need for electricity. This distribution method can be used for other dosed polishing filters also, but tends to be used for more reed beds than other treatment systems due to the fact that reed bed users are also those who typically have a desire to minimise their carbon footprint in the form of electricity inputs. More inspection and maintenance are usually needed for gravity dosing boxes than pumps.

Pros and Cons compared to other polishing filter options

Pros:
- Both soil-based constructed wetlands and horizontal flow gravel reed beds can function without the need for a pumped effluent feed where falls allow.
- Soil-based constructed wetlands provide a wildlife habitat for wetland flora and fauna.
- Wetlands can be attractively designed and landscaped to actively enhance the appearance of the site.
- Like sand filters, vertical-flow reed beds can provide tertiary treatment on a smaller footprint than soil polishing filters. However, where t values are above 20 minutes, a separate distribution area is also needed to disperse the effluent into the ground.

Cons:
- Both types of vertical-flow reed beds listed typically require a pumped distribution or a larger site fall to facilitate a dosing box or bell syphon system.
- Soil-based constructed wetlands are areas of open water with associated risks. Being sewage treatment systems they contain pathogenic microorganisms unless preceded by UV, ozone or other sterilisation unit.
- Horizontal-flow reed beds and soil based constructed wetlands still require disposal of the final effluent via a discharge zone, adding to cost and land take.

Packaged Tertiary Treatment Systems

As discharge standards become increasingly enforced and the wastewater treatment industry responds to the availability of a market for cleaner effluents, package tertiary treatment systems are growing in number and diversity. The different types mentioned in the *Code of Practice* include the following:

- Packaged filters using sand, coconut fibre, or other media – both gravity and pump fed.
- Packaged reed beds – considerably smaller than purpose built reed bed systems.
- Ozone and UV disinfection systems – typically in-line systems that fit as a small component to the piped flow to provide disinfection with minimal space requirements.
- Membrane filtration systems – provide filtration of suspended solids and associated nutrient content as well as many bacteria and other microorganisms present.
- Nutrient removal systems – these include ferric sulphate and aluminium sulphate dosing systems for phosphate removal and systems for nitrogen removal that rely on alternating aerobic and anaerobic activity.

The type of package filter system selected will depend upon the specific requirements of the site, the receiving environment and the preferences of the homeowner. In many instances the tertiary treatment system will be a component within the overall secondary treatment unit rather than a separate element of the treatment setup.

Pros and Cons compared to other polishing filter options

Pros:

- Like sand filters and vertical-flow reed beds, packaged tertiary treatment systems can be more space efficient than soil polishing filters.
- Increasingly, suppliers of package secondary treatment systems are supplying a tertiary treatment alternative within their range, which allows for the one unit to do the tasks of both secondary and tertiary treatment.

Cons:

- Costs are usually higher for package systems, both in terms of capital costs and on-going maintenance costs – however due to the wide range of package treatment options that are lumped together in this category it is important to explore each on its own merits.
- Package treatment systems or combined secondary/tertiary treatment systems still require disposal of the final effluent, necessitating disposal via a distribution area, adding to cost and land take over and above that required for a soil polishing filter, which is in itself both a tertiary polishing system and the disposal route.

CHAPTER 6

Disposal Options

Regardless of the type of treatment system that you put in place, there is still the necessity to dispose of the final effluent. The current *Code of Practice* addresses discharge to groundwater via infiltration in considerable detail, with a number of different disposal options highlighted and described. Discharges to surface waters under licence are identified in the *Code* as a disposal option, but with an acknowledgement that many county councils do not grant licences for discharges to surface waters for domestic situations, thus negating the viability of this as a practical legal solution. Finally discharges to air via evapotranspiration are not mentioned in the *Code*, but neither are they in any way contrary to the aims and objectives of the *Code*, namely the protection of groundwater from domestic sewage discharges. Other disposal routes such as export from the site via sewer or cesspit are also briefly examined in this chapter, along with effluent recycling and volume reduction.

Initial description

For any septic tank or sewage treatment system, there is a practical requirement to dispose of the liquid – regardless of how clean that effluent may be. This section describes the different possible disposal routes and aims to direct you towards finding a solution that will work for your soil, your site and your own personal preferences. Effluent recycling is also examined briefly, but since this only ever works for a certain portion of the effluent it cannot really be termed a disposal option. Likewise, water conservation is listed, not so much because it is a disposal route, but because it assists in the overall limitation of water volume for disposal.

In summary the disposal routes and volume reduction methods described here are:

- Standard soil infiltration, as listed earlier in this book under percolation areas, soil polishing filters and other polishing filters that discharge directly to the ground beneath them such as tertiary treatment vertical flow reed beds and packaged media filter units.
- Direct discharges to surface water.
- Zero discharge willow facilities.
- Filtered discharge to ground surface in low permeability conditions.
- Effluent recycling systems.
- Water conservation.
- Disposal to council sewer.

Note that the section on grey water deals separately with grey water disposal methods, which are somewhat different in character due to the general absence of pathogen contamination that is typically an inevitable part of sewage effluents.

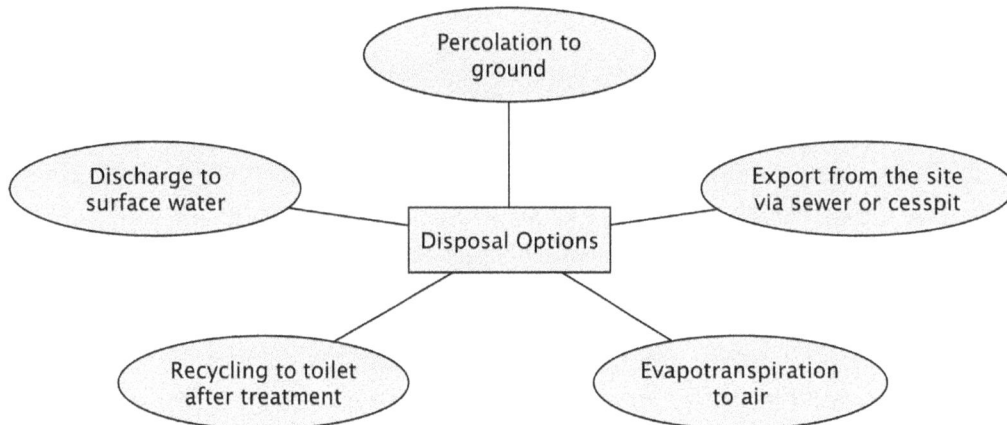

System layout within the site

Disposal options are as varied as treatment options, so each disposal option layout is covered in the relevant section below.

Maintenance requirements for disposal options

The main maintenance requirement for disposal systems is that all preceding treatment components are kept and managed in such a way as to minimise suspended solids and to eliminate sludge loads from carrying through and causing either pollution or clogging of the disposal media, where used.

A further requirement of a very practical nature is to keep the disposal system accessible for inspection. Where disposal methods cannot be identified, the Local Authority septic tank inspection process won't necessarily be digging inspection holes, but it may help you to pass if you have greater knowledge of your overall system. Maintaining easy access to inspection infrastructure such as distribution boxes or pipework will simply make the formal inspection process more straightforward.

6.1   Effluent Disposal Options

Soil Infiltration as per EPA *Code of Practice*

Soil infiltration via a percolation to groundwater can fall into a number of categories as described in the EPA *Code of Practice*:

- *Percolation areas* are the most common form, filtering and disposing of effluent directly from the septic tank. These are typically used in the t value range of 3-50 minutes. The minimum soil depth is 1,500mm below the percolation pipe.
- *Soil polishing filters* are used for disposing of secondary treated effluent to ground. These are typically used where standard percolation areas are unsuitable. Used on t values of 3-75 minutes, or up to t = 90 minutes where p values are between 3 and 75 minutes. Minimum soil depth is 900mm below pipe invert. ('t' applies to the percolation test carried out in the subsoil below the proposed level of the percolation pipe, 'p' to the percolation test carried out in the top soil at ground level.)
- A *distribution area* is used for effluent disposal to ground where tertiary treatment has been used. This applies to soils with a p/t value between 3 and 75 minutes where the depth to groundwater or bedrock is not less than 300mm.

Where suitable ground conditions exist, a **standard percolation area** can be a satisfactory method of disposing of septic tank effluent to groundwater. In suitable soil conditions, further treatment occurs within the percolation area in the biomat layer that accumulates at the base of the percolation trench.

This form of treatment is advantageous because it is a zero energy, low cost method of treatment that can be effective *where suitable soil conditions exist!* Another benefit is that the planners are familiar with its application and are thus more likely to endorse it for planning than some other disposal options. The primary drawback is that not all sites are suitable, and an even distribution of effluent can be difficult to achieve.

(See Appendix 8.1)

Soil polishing filters are described as tertiary treatment options in the EPA *Code of Practice*, but they are also used as the disposal route through which the effluent is discharged to groundwater. These are described in the tertiary treatment section of the EPA *Code of Practice* and this book as *Direct Discharge,* which is used where a treatment system such as a vertical-flow reed bed or media filter discharges to the ground beneath; *Pumped Discharge,* which is used where secondary treated effluent is pumped to the constructed soil polishing area, and *Gravity Discharge,* which is used where secondary treated effluent is routed to a gravity infiltration area with much the same layout as a standard percolation area.

The main advantage of this form of disposal is that due to the use of secondary treatment prior to percolation, it is suitable for soils with a slower percolation rate, extending the upper range from 50 minutes up to 90 minutes where suitable p values are present in the upper strata of soil.

(See Appendices 8.2, 8.3, 8.4)

Distribution areas may be used where effluent has undergone primary settlement, secondary treatment and tertiary polishing. The *Code of Practice* clarification document states that in soils of p/t between 3 and 75 minutes, the area can be smaller again than a polishing filter, and on soils of only 300mm in depth.

(See Appendices 8.5, 8.8, 8.9)

Pros and Cons compared to other disposal options

Pros:
- Soil percolation in soils with good percolation characteristics and suitable depth of unsaturated subsoil is generally the most cost effective option and the easiest to get planning with.
- A good depth of unsaturated soil of good percolation characteristics can offer good filtration of the final effluent, be that secondary or tertiary quality. Compared to discharging the same effluent direct to surface water, soil percolation offers much more environmental protection where ground conditions are suitable.

Cons:
- Not all sites have good percolation characteristics, i.e. t values between 3 and 90 minutes.
- Not all sites have a sufficient depth of unsaturated overburden, i.e. subsoil that is above groundwater level and above the bedrock.

Sometimes it is possible to plant over percolation areas to encourage evapotranspiration, or nutrient uptake for recycling as mulch and a compost ingredient. On a cautionary note, the EPA *Code of Practice* recommends that percolation areas be grassed, which, although boring and not necessarily a very productive use of garden space, is recommended to ensure that soil movement down into the percolation area is minimised and the danger of clogging pipes with root growth is eliminated.

Sometimes willows are used to dry out percolation areas, but this should be planned with care to protect the pipes from the invasive root growth that can completely fill the pipes to the point of blocking them. Where willows are used in zero discharge facilities, a carefully designed distribution pipe setup is used to prevent this. The best place for willows is well down gradient from percolation areas to filter groundwater and to soak up surface runoff.

Planting with comfrey has been cited as a way to draw N, P and K out of the water in the percolation area for recycling to the compost heap or as direct mulch in the garden. This should only be used with careful planning to avoid pipe clogging by roots, and should only be carried out where the materials reaching the percolation area are carefully controlled so that you don't end up with a comfrey plot full of chemicals from cosmetics, cleaners and detergents.

Direct Surface Water Discharges

A direct discharge to a drain, stream or river is a legally acceptable option on poorly drained sites if both the degree of treatment and the volume and health of the receiving water are sufficient. As a minimum a 20/30 standard (mg/l for BOD and suspended solids) is required with a minimum dilution in the receiving watercourse of 1:8. This water quality standard is easily achieved with a constructed wetland or other secondary treatment system, but the dilution is not always available. Even with sufficient dilution, there will still be bacteriological activity within the final effluent, so the use of the receiving water downstream is a consideration.

More recent legislation such as the Surface Water Regulations (2009) stipulate more stringent guidelines than were previously applied, making discharges to surface waters more difficult to achieve from a treatment perspective. In addition, many councils have a policy of not granting pollution discharge licences to homeowners, essentially removing this as a viable disposal route. Nonetheless, in practical terms, a surface discharge is often what is taking place where houses are already built on sites without suitable percolation. See 'disposal to ground', at the end of this section for more details.

Pros and Cons compared to other disposal options

Pros:
- Can provide a mode of effluent disposal where soil characteristics do not permit a discharge to ground.
- Can provide a safe way to dispose of effluent where the treatment standard prior to discharge is of the very highest quality.

Cons:
- High potential for water pollution if the treatment system is not very carefully monitored and maintained – or protected by a large constructed wetland or reed bed after secondary treatment to provide buffering for any spike in hydraulic or pollution loadings.
- Legally requires a discharge licence under the 1977 and 1990 Water Pollution Acts. This typically requires, *inter alia*, regular monitoring of effluent and reporting to the relevant local authority, along with an annual maintenance fee payable to that authority.
- Many local authorities do not grant such licences for domestic effluents, ruling this option out.
- Pre-treatment requirements are more stringent and consequently can be more costly to implement.

Evapotranspiration and Disposal to Air

The most effective evapotranspiration option in this climate is the use of quick growing cultivars of *Salix viminalis* willow. Other trees such as red alder and eucalyptus also have a very quick growth rate and high water throughput, but are not as suitable. Alder is a nitrogen fixing plant and thus not as suited to nitrate uptake from sewage. Eucalyptus is frost tender, and even fully mature trees can be killed off in exceptionally harsh winters, as recent years have shown.

Zero Discharge Willow Facilities were developed in Denmark by Peder Gregersen of the Centre for Recycling in 1996 as a way to evaporate 100% of the effluent pumped into them. The idea was a development of previous use of willows for sewage treatment in combination with the emerging willow cultivars being bred for the emerging Scandinavian biomass industry. The result is a very effective sewage treatment/disposal method which can double as a source of biomass timber for burning, and which also has benefits for wildlife and makes a visually attractive landscaped part of the garden. See the treatment options section for more details on willow facilities.

(See Appendices 8.11, 8.12)

Another method of disposal to air involves incineration toilets (I kid you not – they do exist), but this still leaves grey water to be disposed of. See the treatment options section for more details, but in summary, this system is not going to win you the environmental home of the year award.

Treatment wetlands also have an evapotranspiration element, and may be able to achieve a zero discharge with careful design and siting, particularly during summer months.

Pros and Cons of willow facilities compared to other disposal options

Pros:

- Willow facilities dispose of all effluent to air, so there is no discharge to ground or to surface water. This guarantees that there is no pollution risk to aquifers, wells, streams or other water bodies.
- Because there is no discharge, there is no surface water discharge licence requirement.
- Willow facilities are the only treatment system/disposal option that will yield more energy in the form of firewood or wood chips than they consume in their construction and maintenance.

Cons:

- They can be a costly system when compared to a septic tank and percolation system. That said they can become quite economical when compared to a mechanical secondary treatment system followed by a wetland or reed bed, accompanied by an on-going water pollution discharge licence.

Disposal to Ground – On Low Permeability Soils

Like it or not, some sites do not have soil of sufficient permeability to facilitate infiltration, do not have sufficient access to streams with sufficient dilution rates for a surface water discharge, do not have the space or budget for a zero discharge willow facility and/or do not have access to a council sewer. However, that won't necessarily stop you wanting to live in a particular part of the country if you happen to call it home. There are also many houses already built where discharge conditions are unsatisfactory and options are limited.

Prior to the *Code of Practice* a number of options were in use and accepted to a greater or lesser degree by many councils. These essentially involved limited percolation on poor soils or informal surface discharges following effluent treatment and/or tertiary polishing by constructed wetlands, reed beds or raised percolation areas.

Although these are not typically acceptable to local authorities for greenfield sites, they may be used to provide additional treatment on sites that are currently causing pollution. These options may also be used as a practical solution for a new dwelling where the council choose to accept them.

A soil polishing filter surrounded by a perimeter drain typically routes excess filtered effluent to an adjacent surface drain or stream. This is for effluent that has been treated in a secondary treatment system and pumped to the raised mound for tertiary polishing.

A willow planted soil mound is a variation on this method, in which willows are used as a filter plant to take up nitrogen, phosphorus and liquid volume. This variation has the advantage of minimising or eliminating summer discharges to small watercourses when the dilution rate is low or non-existent.

A reed bed, willow bed combination was the recommended method by some counties prior to the *Code of Practice* introduction. This method treated septic tank effluent in a gravel reed bed system and routed this effluent into a constructed willow bed area to provide uptake of N, P and seasonally variable amounts of effluent volume. As per the willow planted soil mound, this method had the advantage of limiting summer discharges when dilution was low in receiving waters.

The percolating pond or marsh is a method that relies upon limited percolation to subsoil, some willow evapotranspiration and infiltration in to topsoil in the perimeter of the pond or marsh for excess liquid. This method has the advantage that although it is used in soils with t values over 90 minutes, it can be effective at disposing of well-treated effluent into the ground without recourse to surface discharge at any time of year.

Lateral willow filters are planted areas of *c.*300mm depth of suitable soil, *c.3*-5m in width between a pond or treatment wetland marsh area and a receiving drain or stream. Planted with willow trees and kept coppiced as per a standard zero discharge willow facility, these lateral filters can provide additional uptake of nutrients and liquid volume on sites with zero soakage. They can work well with a gravity fall after a standard constructed wetland or reed bed, or as a hybrid with a percolating pond system.

Planting of existing surface ponding or seepage can be carried out to assist in the overall resolution of problem sites. Surface ponding or seepage is not a recommended disposal route, but is not uncommon nonetheless. This occurs where the effluent treatment system discharges directly to ground surface, or where percolation areas cannot discharge vertically to ground, and well up at ground level. Where this occurs, an improvement may sometimes be achieved by planting the area with constructed wetland plants or willows as appropriate. Fence the area effected to contain the effluent and treat the area as one would a surface-flow constructed wetland system in terms of health and safety precautions.

Planting of receiving drains as buffer zones. Many drains in areas with poor soakage already receive septic tank effluent from adjacent houses, sometimes to the point of being open sewers. It is worth noting however that planted drains will function in the same way as a constructed wetland system and start to clean up the effluent being discharged into them.

System layout within the site

For formal systems such as raised mounds, willow planted mounds, percolating ponds etc. all *Code of Practice* minimum distances should be adhered to. With informal areas such as the planting of existing ponding or seepage, or planting of buffer zones these minimum distances won't necessarily be possible. However it's better to have a planted channel (buffer zone) that is too close to a drain, than just a plastic pipe with no treatment potential.

(See Appendices 8.6, 8.7, 8.10)

Pros and Cons compared to other disposal options

Pros:
- Where discharge options are limited and the house is already in place, these can be practical disposal methods that can yield significant environmental improvement over a piped discharge from a septic tank or treatment system to a drain.
- These are relatively low budget solutions where financial resources may not be available for a zero discharge willow facility.
- For the willow planted options, discharges tend to occur mostly in winter when the dilution rates in the receiving drain will be highest; and then lowest or absent in summer when the receiving drain may be dry. Hence there is a practical environmental benefit.

Cons:
- These options are not covered in the EPA *Code of Practice* for greenfield sites.
- Willow planted options can be difficult for councils to legislate for since there may or may not be a surface water discharge at any given time.
- Contaminants may carry through to the receiving drain or ground surface at certain times of the year.

6.2   Effluent Volume Reduction

Although effluent volume reduction is not a disposal route – it does offer a viable way to reduce the overall volume that requires disposal and as such offers a way to minimise the required surface area for disposal options to soil and reduces the dilution volumes required for discharges of effluents to surface waters.

Effluent (and Grey water) Recycling

Recycling is an option for effluent from some sewage treatment systems, particularly those systems that produce a particularly high quality effluent. This option does not necessarily allow construction on sites with a t value >90 minutes, which still won't satisfy the *Code of Practice*. Nonetheless where soils can physically receive water for disposal to ground, even t>90, where the effluent has been tertiary treated, this may be a practical option, which may be permitted under the heading of environmental gain. Where effluent recycling comes in useful is where the site size or the soil hydraulic assimilative capacity (the ability to absorb the water) are limited, and where the reduction of the total water volume by about 50% would make the difference between being able to provide adequate treatment and not.

There are arguments for and against effluent recycling. On the plus side it should keep usage of piped potable water lower than would otherwise be the case, and it should minimise the effluent load on the receiving environment. On the other hand, additional electricity is needed to achieve the levels of treatment needed; pumping is required to recirculate the water; and additional infrastructure needed to store it. It is probable that changing technologies will improve the environmental footprint of water recycling units, but doing a careful assessment of environmental costs and benefits is recommended before jumping in.

Direct recycling is another option, such as using bath water for irrigation, combined with the use of compost toilets where a high environmental performance or low water usage is desired. Take care with reusing bath and shower water in your garden that all shampoos and soaps etc. are literally good enough to eat! Health shops offer a good range of eco-friendly options in this regard.

Pros and Cons of reduction compared to other disposal options

Pros:
- Can be used to minimise effluent discharges.
- Can be used to minimise water consumption and water charges.

Cons:

- May have a relatively small water saving contribution in practice, depending on system type and the way you use it.
- Recycling typically requires mechanical treatment and an additional pumped return to the dwelling, which add to electricity consumption.
- Requires additional plumbing infrastructure.

Water Conservation

Although not a disposal option *per se*, water conservation is an obvious factor in reducing your overall effluent discharge volume. Although not the case for all sewage treatment systems, all the natural, planted systems such as reed beds, wetlands and willow systems function more effectively if there is a lower throughput. Rather than finding ways to get rid of so much water into poor soils, cutting back on its use can have the direct effect of improving final effluent quality as well as reducing the area requirements for disposal via percolation.

In terms of water conservation in the home, there is a lot that can be done to minimise the overall volume of effluent reaching your percolation area, stream or willow facility. The only effluent system that I am currently aware of that takes account of water savings is the zero discharge willow facility, since the design of these systems is highly dependent on user lifestyle in this respect. Hence, the smaller the volume, the bigger the cost savings for the homeowner. That said, most treatment and disposal systems benefit from reduced throughput.

Simple steps such as turning off taps, fixing leaks and putting a litre milk jug of water in your toilet cistern can save surprisingly high volumes each year. Dual flush toilets and water saving showers and appliances can up this amount even more. Until metered water charges are introduced, water saving will remain in the domain of the environmentally aware and of businesses (including our schools, farms and offices) where metering is already in place.

There is a wealth of water saving resources available on line for those who wish to pursue it further. Refer to the web references at the end of the book for more details.

6.3   Export Via Sewer or Cesspit

Although not the primary focus of this publication by any means, export of the effluent from the site is sometimes an option and merits a brief mention. If you can connect to a public sewer or group sewage treatment scheme, this is an easy way to export your problem, so to speak. You pay for the service of having the effluent removed, treated (hopefully to an adequate standard) and disposed of to a watercourse.

Export to Sewer

Sewers are generally available in all urban areas in Ireland. Where sewers do not exist, but there is a high housing density, it may be feasible to organise a group sewage scheme to remove and treat all of the effluent in a given area where soil conditions make on-site treatment difficult or expensive.

Pros and Cons compared to other disposal options

Pros:
- You can hand over responsibility for both treatment and maintenance to the local council or group scheme management company.
- Addition of one more toilet to the local sewage treatment plant probably won't make that much difference to the river it's discharging into, whereas taking out the last few poorly performing percolation areas from a groundwater aquifer or a small stream or drain may make a significant improvement to them.

Cons:
- Connection costs for public or group sewage can be quite expensive, particularly where a long pipe-run is required to connect to the nearest sewer.
- Long-term costs for public sewage treatment are unknown, however if and when sewage treatment charges are introduced it is likely that they will be relatively modest.

Cesspit Storage

Cesspits are not currently an accepted option in Ireland, but due to the challenges posed by the current septic tank inspection process, it is likely that this will be changed. Although not an ideal option, storage of your sewage in a large tank and export by tanker may be a viable option where other treatment approaches are deemed to be unacceptable. Essentially any site that causes pollution and where the options set out in the treatment section of this book either will not work for some reason (e.g. site too small or too close to a lake etc.) or are deemed to be unacceptable (e.g. dry toilets are not for everyone, discharge licences may not be granted by the council etc.) then storage and export by tanker is an obvious, if expensive and somewhat cumbersome, solution.

Pros and Cons compared to other disposal options

Pros:
- You can achieve protection of the local environment on situations which may otherwise cause pollution. This is really only valid for legacy sites where the house already stands and pollution is already occurring. It should not be used as a viable solution for unsuitable green-field sites due to on-going maintenance inputs and costs.
- Ease of immediate installation. In some EU countries, a temporary cesspit in the form of a plastic tank can be simply dropped on site and sewage can be pumped up into it from a sump or septic tank.
- Low capital costs compared to an extensive treatment system.

Cons:
- Generally unacceptable legally, although the septic tank inspection process may change that.
- On going costs for removal of effluent to a local sewage treatment plant. Costs may be very significant unless extensive water conservation measures are adopted and adhered to.
- Regular tanker access needed to draw effluent away from the site.
- Large plastic standing tanks, where used, can be visually obtrusive.

CHAPTER 7

Buffer Zones

Buffer zones are areas of land that act as a filter between potential or existing sources of pollution and the receiving environment – typically a river, stream or lake. Buffer zones are most common in an agricultural setting where they help to filter out soil, fertilisers or biocides from field runoff. At the surface of the ground, buffer zone vegetation filters physical debris from runoff water and slows down the flow to enable soil and silt to settle. Within the soil, roots – particularly tree roots – take up nutrients as the groundwater moves underground towards watercourses. Another relatively obvious effect of buffer zones in an agricultural context is that by excluding livestock from stream and drain banks, they protect the watercourse from direct pollution by ground disturbance and animals defecating and urinating in the water directly. The overall effect of buffer zones is to protect watercourses and groundwater from dispersed pollution (as opposed to 'point source' pollution which is from a single, identifiable source, which is most readily dealt with by treatment at that source).

The two areas that come to mind in the context of septic tank options and alternatives is where (a) there is existing problem where polluted water enters your site from an adjacent field, drain or stream and (b) where you want to add an additional buffer area for dealing with residual pollution from your own sewage treatment system.

7.1 Dealing with Existing Pollution from Outside Your Site

The majority of sites do not suffer from surface pollution from neighbouring septic tanks or farms, but where it occurs it can render an area of the garden anywhere from unpleasant to a health hazard. Usually the best option for dealing with a source of pollution is to address it at source with whoever is causing it. However, this is not always an easy route to follow for any number of reasons. Where the pollution source is not possible to address at its source, a buffer zone can be employed to at least clean up the water to contain and filter it within your own site.

Four buffer zone layout options are listed below, but in reality, buffer zones do not need to fit into strict categories to be effective. Nonetheless, careful design and implementation are recommended to ensure that they work to the best of their ability.

- Conventional buffer strips
- Linear buffer wetland channels
- Linear buffer willow channels
- Willow strips for groundwater protection

Conventional Buffer Strips

Conventional buffer strips comprise an exclusion zone of land running in a long strip between the source of pollution and the environment they are designed to protect. These are typically planted with grass, wetland plants where appropriate, or trees. Typically used in agricultural situations, at their simplest these consist of a fence, which protects the buffer zone from livestock and allows natural regeneration to take place. Conventional buffer zones are not necessarily the most space efficient option for a home garden since they do not contain the polluted water, but simply allow it to pass through. These buffer zones are also better where the pollution is from a diffuse source such as a field rather than a single point of entry such as a drain flowing through your site.

Linear Buffer Wetland Channels

A linear buffer wetland channel is essentially a planted channel that allows water to benefit from wetland filtration as it passes along its course. An alternative title could easily be 'planted drain'. When drains are cleared they allow water to pass easily along their course, moving more swiftly in wet weather when the runoff from fields is at its highest and providing relatively little scope for settlement before reaching rivers downstream. By contrast, as soon as a drain is planted with tall wetland plants (or left undredged if plants already exist), the water slows down, soil and debris settle out, nutrients are taken up by growing plants and a host of physical, chemical and biological processes take place to clean up dirty water.

Essentially a linear buffer filter wetland behaves like a constructed wetland system, except that the design layout is dictated more by land availability than pollution removal requirements. Naturally the higher the pollution load, the bigger the buffer area should be, but in practice, not all gardens have the space to filter other people's pollution problems, so these buffer areas provide a way to do that cost-effectively and provide an attractive and well contained wetland habitat into the bargain.

Linear Buffer Willow Channels

These function in much the same way as linear buffer wetland channels, except that rather than planted with wetland vegetation they are planted with quick growing cultivars of willow. These can be particularly effective where the ground gets waterlogged, since the willows will take up a large volume of water during the growing season. In winter, the evapotranspiration is considerably less, although not absent. However the mat of roots that forms beneath the soil can make the ground more solid, and the presence of the willow trees can provide a framework to the garden that allows a wet area to at least be beautiful during the dormant season, rather than expecting a lawn mower to battle through soft ground just to keep up an area of open lawn on unsuitable ground.

Willow channels have the advantage over wetland channels in that their roots extend deep into the soil and can address suspected or known groundwater pollution. Their drawback is that surface water is filtered better in a drain of wetland plants than willow stems.

For dealing with both surface pollution and groundwater pollution, a wetland and a willow component can be employed as a buffer area one after the other. You can also plant a channel with both willows and wetland plants, but care needs to be taken to prevent the willows simply overcrowding the wetland plants beneath.

Willow Strips for Groundwater Protection

Instead of planting willows in a defined channel, the trees may also be planted as a strip or curtain between the pollution source and receiving watercourse. This approach follows the lines of a conventional buffer zone, but is intended specifically for groundwater rather than surface runoff, as well as having the intention of drying the soil where the trees are planted.

This approach can be particularly effective where there are signs of pollution in adjacent drains or streams or where a raised percolation area has been installed and surface ponding is apparent. The willows may be planted at the drain edge; or below the mound of a raised percolation area, to take up nutrients as the water moves past the roots.

Note that where willows are planted near percolation pipes they can grow right into the pipe perforations and lead to blockages. Thus careful design and implementation is needed in order to avoid creating further problems on the site. Willow roots can easily spread to 10m and fill pipes if left unchecked. If they are coppiced they tend to be less invasive. Willow strips can be an excellent way to mop up current or historic pollution even at a distance from the source, so pipe clogging may not be a problem at all.

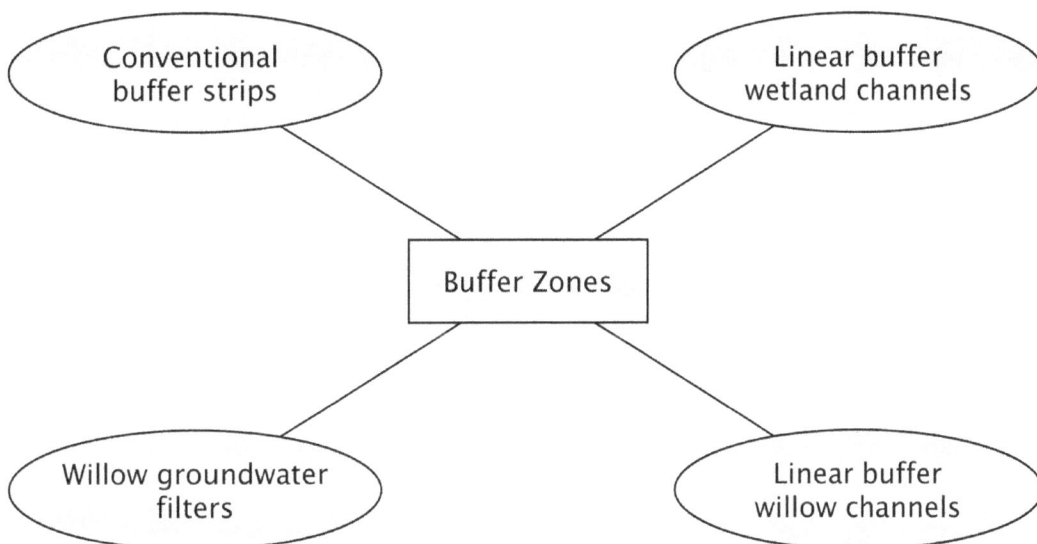

7.2   Dealing with Residual Pollution from Within Your Site

If you are the one causing the pollution, then the starting point must be back at the beginning of this book. Your septic tank or treatment system should work; any filter wetlands etc. should be large enough, well maintained and working well; and the percolation area should function correctly, or the surface discharge point clearly identified and maintained for examination by the council if needed.

However, sometimes there are situations where a bit more filtration would benefit the receiving environment, and in such cases, the use of buffer zones can be a valuable addition to the site. In this instance, any of the above systems can be effective, each designed to suit the site conditions and layout. Also, there is the potential to use an additional buffer type that I refer to as an emergency response wetland, intended both for buffering existing effluent, and acting as a safeguard against possible treatment system failure.

Emergency Response Wetlands

Most package sewage treatment systems rely on air blowers working 24 hours a day for 7 days a week for their continued effectiveness. Zero discharge willow facilities, raised percolation areas, and any percolation system or treatment unit up-gradient of the septic tank rely upon electricity to deliver the effluent to the treatment area. Pumped delivery of effluent requires considerably less electricity input than continual air pumping, however all of these systems rely on the continuous supply of electricity that we have come to assume will always be present.

What would happen if, in the morning, that electricity supply suddenly lost its reliability? Ireland relies on imports for about 90% of our energy usage. Oil wars, peak oil, climate change policy or natural disasters could all interrupt the reliable availability of fossil energy, and electricity as a direct result. Even taken in a somewhat less catastrophic view, rising energy prices already have some people turning off their treatment systems (this is not recommended as a way to save energy! Start by buying a system that doesn't need electricity inputs in the first place).

So, back to the question of energy supply... Without electricity inputs, treatment systems discharging to surface waters would start to cause pollution of rivers, lakes and streams. Treatment systems discharging to polishing filters would quickly begin to discharge a higher solids load, leading to clogging and possible surface ponding or groundwater pollution. Septic tanks that rely upon pumping to a treatment system, wetland or willow facility would begin to overflow or back up.

All in all, the wastewater treatment capacity of the country would be somewhat compromised, to put it mildly. Now, who do you think will actually be too worried about their treatment system in such a scenario? Considering the level of importance that sewage treatment generally takes in terms of the overall house design process, my suspicion is that sewage treatment won't be at the forefront of our minds as we try to grapple with a sudden wrench from our energy source of choice.

Enter the Emergency Response Wetland. Here is an official proposal that all sewage treatment systems and all septic tanks should have a back-up plan to cater for interruptions in electricity supply over and above the reserve capacity in the tank or treatment system. This wetland consists of an area of the site that is built to be lower than the tank, that will be a natural overflow route in the event of prolonged power failure. Any water that overflows or receives inadequate treatment can be routed through this purpose built wetland area to filter water and provide protection to the receiving environment without the need for electricity.

The system can be used as a clean wetland wildlife area, or bog garden in the intervening time, providing a wetland habitat to plants, birds and dragonflies. There are many parts of the country where poor site drainage would benefit from such an additional wetland space – even with all the preceding treatment components working perfectly.

One feature of the Emergency Response Wetland is that being a buffer zone, it is not bound by *Code of Practice* minimum distance requirements for sewage treatment systems. Certainly it improves water quality, but only as an add-on to a working system, or insurance against possible malfunctions.

STEPPING INTO ACTION

From Decision to Implementation

CHAPTER 8

Planning, Implementation, Maintenance

Having selected a sewage treatment system the next step is to start the process of obtaining planning permission for works on site with a view to putting it in. Maintenance is also included in this section, since proper care of the system you choose is vital to its on-going success.

Each system is very different, but this chapter gives a quick summary of the steps involved in bringing your selected system into being.

8.1   Planning Requirements
When Do I Need Planning and How Do I Apply?

Planning permission is usually required to install or significantly change any sewage treatment system, even if it means that your overall environmental impact will be reduced. The easiest and cheapest time to apply is when you are applying for any other works – be they extensions or alterations to your home, or construction of a new-build project. However if the only thing you need to apply for is your treatment system, then if your site conditions are favourable, it should be a fairly straightforward process. The majority of the costs will be in the preparation of the planning application and in the site works themselves. That said if your site conditions are favourable, you probably aren't going to need to upgrade your system. It is worth noting that the *Code of Practice* really only requires good percolation on new sites. Where the house is already in place, environmental gain is taken into account, even if it is achieved with methods that do not otherwise comply with the *Code*.

Septic tank inspection process aside; it is difficult for a local authority to demand that a house install or upgrade their sewage treatment system unless water pollution is clearly identified locally as being a problem. This is the case even in situations where local ground conditions are known to be unsuitable for septic tank and percolation systems. However, one route that local authorities can and do adopt in known problem areas is to insist on a new treatment system where any planning applications are lodged for extensions or alterations to the house itself.

In this context, bear in mind that if you are already living in an area known to have unsuitable soil characteristics and you wish to carry out any renovations on your house, then you will probably need to address your sewage treatment system as well as your new bay windows. That said, you wouldn't have read this far if you didn't want to deal with your sewage treatment anyway, so it's a bit of a moot point.

The following general information will usually be required for any planning application for a sewage treatment system:

- The site assessment report carried out by your site assessor. This shows the percolation test and trial pit results, as well as other points of relevance to the site characterisation.
- A description of the proposed system, outlining why it is needed and what it entails. This should include the expected discharge quality or reference to the relevant section of the *Code of Practice*.
- Drawings are usually required, with a site layout map showing where the new system is to be located within the site. Sometimes a topographical map is useful, or required by planning, showing the contours on the site.
- Details of the proposed discharge route, and how this conforms to the EPA *Code of Practice* or an explanation as to why the *Code* route is not being followed, what alternative is proposed and how it provides adequate protection of the environment. Note that if you don't follow the *Code* guidelines, then getting planning approval can be more difficult, and certainly not guaranteed.
- Standard planning requirements such as a site location map; copy of the site notice and newspaper notice announcing the proposals (both of which need to be made public within a specific timeframe around the application date, so read this section of the planning application material very carefully); and 6 copies of all maps, drawings and reports submitted.
- Such information pertaining to surface discharges, archaeology, ecological assessments and the like as may be relevant to the site.

The most straightforward way to apply for planning is to employ an architect, engineer or planning consultant to carry out the application process. However if you wish to submit for planning yourself, all of the relevant information on how to apply is usually clearly set out in the planning section of your local County Council website.

When Do I Not Need Planning?

There are circumstances where planning permission is not required for changes, alterations or upgrades to a sewage treatment system, as follows:

- Typically upgrades to existing treatment systems or septic tanks and percolation areas do not require additional planning permission where the existing footprint and technology remains unaltered. Repair and maintenance are not planning issues, whereas moving tanks or percolation areas to a different location, or changing from one treatment system type to another, generally does require either planning or consent from the local authority environment or planning departments.
- Where you are served with a Section 12 Pollution Notice by the local authority. If you are served with a Section 12 Notice for causing water pollution then it is understood that there is an existing pollution problem that needs to be addressed as a matter of urgency. The usual planning process does not apply in this case, and all plans and particulars are submitted directly to the environment department of your local authority, who make the decision as to the suitability of your proposals.
- If you have already received planning permission for a particular sewage treatment system and wish to change this to an alternative system then you may make a written submission to the local authority environment department explaining why

you wish to change it and outlining your preferred system in detail. The environment department may then permit an amendment to your original planning application. Keep careful copies of all correspondence so that you can demonstrate planning compliance should the need arise.

- If you have failed a formal Local Authority septic tank inspection, then you will not need planning permission to implement the recommended changes, and you may also be eligible for a grant to assist with the works.
- Garden ponds, bog gardens, buffer zones and the like do not generally require planning permission. These may feasibly be irrigated with grey water, but not serve as treatment systems *per se*.

Other Planning Considerations

There are considerations other than straightforward planning permission, including for example discharge licences to surface waters, proximity to sites of ecological importance and proximity to significant archaeological features.

Where a discharge to surface waters (i.e. a drain, a stream, a lake or the sea etc.) is proposed, then legally you will need to apply for a discharge licence from the local authority. This is a separate process from the planning application process, so it is theoretically possible to get planning for your house; build it; and then not have your discharge licence approved and therefore not legally be permitted to dispose of any water from the house. To avoid this, always apply for your discharge licence in parallel with your planning application process where a surface discharge is needed. Many local authorities do not grant pollution licences to homeowners, which means in practice that discharges to surfaces waters are often not an option. A local engineer or environmental consultant should be able to prepare a discharge licence application. As with planning, there is no guarantee that the licence will be granted after you go through the application process.

Sites of ecological importance such as Natural Heritage Areas (NHAs), Special Areas of Conservation (SACs) and Special Protection Areas (SPAs) require protection above and beyond that afforded elsewhere. Thus if you are located within an NHA, SAC or SPA, or if your effluent will discharge to a river or stream, whether via a percolation area or a direct discharge to that stream, then an assessment of the ecological impact is often required as part of the planning process. An Appropriate Assessment is a requirement under the Habitats Directive to provide protection to SACs and SPAs. An AA is typically carried out by a local ecologist or environmental consultant.

Archaeological features near a proposed sewage treatment system may raise obstacles to the type of system selected, or whether you can install any system in that area. Smaller footprint systems may be preferable if an archaeological feature limits the overall area available for site works. A local archaeologist will typically be needed if such a feature is present.

With all of these circumstances, initial discussions with your architect or engineer should identify any site-specific restrictions. Preplanning discussions with the Local Authority will also be helpful in identifying any potential limitations or restrictions on the site.

As of March 2014, a new building control process was initiated in Ireland which can add time and expense to the planning process. Discuss this with your architect or engineer early in the process to see how it may effect your project.

8.2　Starting Work

Before you ever begin, it's a good idea to have an idea of what it will all cost. The following list of items should cover most of the costs in a given project, so pick up the phone, talk with the relevant consultants, suppliers and installers and start totting up the estimates and quotes you get for each item:

- System design report and drawings from site assessor, environmental consultant or engineer.
- Architect or engineer planning application process fees.
- Planning fees.
- Consultant's fees for any requests for further information (RFIs) from the council, or for resubmissions, appeals, or amendments to the planning file.
- Treatment system capital cost or cost of materials.
- Construction, installation, connection, commissioning. Consider insurances, site safety processes and other site works as may be required to actually do the job. Are there electricity cables, gas lines, water mains etc. that may need to be moved or protected? Will subsoil need to be removed from the site, imported, or spread around, and how much will this cost? Electrician costs need to be considered. Plumbing to the house itself may be an additional cost, or toilet replacement if you are putting in a dual flush toilet to conserve water or to reduce the load to a willow facility. Is there a construction follow-up needed (such as for willow facilities, which need work carried out in the spring of the year following initial construction to finalise embankment detail)?
- Landscaping, fencing, tidying up. Not usually a major consideration at the beginning of a project, but essential to consider and budget for if you want to get your garden back the way you want it at the end.
- Maintenance and running costs for the system you plan to use, be that for willow harvesting, secondary treatment system checks and pump replacements, electricity costs, desludging costs, discharge licence fees and wastewater analysis costs etc. Itemise them all for your selected system and get prices for each item.

If you are used to getting a construction project done, then supervising it and organising it yourself is usually the most cost effective way to proceed. If you'd rather not have that hassle, then consider who would be the best person to manage the job. A site foreman is often the easiest if you have a bigger building project on the go already. A local engineer, site assessor or environmental consultant may be able to do that work. Alternatively work hand in hand with whoever will be doing the digger work in the case of a constructed system, or the supplier in the case of an installed system, and ask them to itemise all the costs, and how they're going to do the job and whether there is anything that needs doing that they're not planning to do – so that you know about it in advance and can appoint an alternative contractor for that project element (such as final landscaping for example).

So what's actually involved in doing the work? Well, depending on the system you are putting in, you will have slightly different elements. However the following list gives a good overall work plan summary:

1. Design the system and procure planning permission. Designs can usually be drawn up for both planning and construction at the same time. If you are going for a mechanical system then you usually still need a percolation area layout – even if only shown on your architect's site layout plan.
2. Remove topsoil from the work area. Stockpile for reuse in the system (if a free water

constructed wetland or willow facility) or for reuse in the garden if an alternative system.

3. Excavate the subsoil as necessary for your settlement tanks, reed bed, willow facility or whatever system you have selected. Subsoil may need to be distributed elsewhere on site or exported from the site altogether.

4. Install your percolation trench work, tanks, reed bed liner and gravel, willow facility liner and soil etc. as required by the particular system being used. Build any outlet flow control units, inspection manholes, sampling access manholes as may be needed.

5. Complete the relevant pipework connections between the system components; except the final connection from the house to the septic tank or from the tank to the treatment system. This is to ensure that no sewage enters the system until all planting of wetlands, completion of pump or electrical works etc. have been completed in full.

6. Plant wetlands, reed beds, willow facilities. Install any mechanical treatment system components as may be necessary. Finalise outdoor electric wiring.

7. Connect to the sewer pipe from the house only when wetland plants are established, all mechanical components are in and tested, willows ready etc.

8. Check the whole system and get a letter of compliance from the supplier, installer or designer.

9. Follow-up maintenance is important and will vary from system to system. Some systems require more inspection and maintenance in year 1 than in subsequent years. Willow facilities are a case in point where they often need follow-up construction at the end of the first season's growth. But regardless of which system you have – *all treatment systems require maintenance*! Even soil-based constructed wetland systems still need to have the septic tank emptied as appropriate.

That should cover most systems, but talk the project through in full with *all* members of the design and construction team before breaking a sod or writing a cheque. Happy digging.

8.3   Maintenance Matters

The title says it all. Maintenance does matter. Even if you are only empting your septic tank, it is important that it is carried out in good time to avoid sludge carrying over into your percolation area or gravel reed bed. It would be a bit galling to have to do expensive replacement works to your new sand filter or zero discharge willow facility just because you didn't empty your treatment system, settling chamber or septic tank in time. Remember that a foot of water from the top of a heavily sludged tank to the outlet pipe may stop the sludge itself from being drawn into your percolation area or filter, but it won't provide adequate settlement time for the new solids entering the tank every day. Thus your suspended solids load on the percolation area will be a lot higher than it should be.

Regular desludging will presumably be needed under the new Local Authority inspection process. The EPA *Code of Practice* currently recommends annual desludging as a minimum. An alternative desludging regime was set out by Cavan County Council before even the European Court Judgement condemned Ireland as being distinctly behind the times in this regard. Because of the Cavan County Council inspection guidelines, it was the only council specifically exempt from the European Court Judgement that started the whole septic tank inspection process in the first place. Rather sensibly the Cavan guidelines recommended inspection and desludging frequencies that were a reflection of tank size relative to the household population rather than a blanket one-year frequency. Although the Cavan bye-laws are no longer in use, the table is repeated here as an example of how tank size and population impact on the anticipated sludge accumulation rates in a septic tank or treatment system:

"Sewage sludge shall be removed from septic tank systems at intervals not exceeding that specified in Table 1, or as determined by inspection, so as to allow the system to continue to operate effectively."

Table units indicate interval in years between desludging times. For 6 people with a 3400 litre tank, desludging should be once per year. If the same tank is used for only 2 people, the frequency could feasibly drop to every 4 years without compromising the effectiveness of the tank.

Tank Size (Litres)	Household Size (Number of People)									
	1	2	3	4	5	6	7	8	9	10
2,250	5.8	2.6	1.5	1	0.7	0.4	0.3	0.3	0.1	
3,400	9.1	4.2	2.6	1.8	1.3	1	0.7	0.6	0.4	0.3
4,500	12.4	5.9	3.7	2.6	2	1.5	1.2	1	0.8	0.7
5,700	15.6	7.5	4.8	3.4	2.6	2	1.7	1.4	1.2	1
6,800	18.9	9.1	5.9	4.2	3.3	2.6	2.1	1.8	1.5	1.3
8,000	22.1	10.7	6.9	5	3.9	3.1	2.6	2.2	1.9	1.6

Remember that septic tank or sewage treatment sludges intended for land spreading to agriculture or disposal to council facility are bound by the Waste Management Regulations, so be sure that whoever empties your septic tank is licensed to do so and can give you a certificate of completion that you can keep filed away for a formal inspection if needed.

Beyond desludging, all sewage treatment systems require different maintenance inputs. It is best to read the relevant section of Chapter 6, refer to the EPA *Code of Practice* and check supplier information for additional information.

Even though the formal septic tank inspection process has started in many counties (at the time of writing) exactly how it performs and how rigorous it will be remains to be seen. However it is recommended that you keep a file recording all maintenance works that you carry out, including pump changes or services, tank desludging, and any self-inspection or supplier-inspection of your tank or treatment system. This information may be helpful for the formal inspection process to verify dates and compliance.

If you are carrying out any works, take photographs and preferably work on the recommendations of a registered site assessor, engineer or environmental consultant. In some counties, if a site assessor didn't sign off the work it may need to be redone to verify compliance with the *Code of Practice*.

On-going maintenance can be time consuming and costly, but please remember that not carrying it out can cost you a lot more when the time comes to replace an entire system as result of sludge congestion or a water pollution event.

Closing Words

What I have endeavoured to do in these pages is present a range of treatment systems and management options side by side, whether or not they fall within the current EPA *Code of Practice*. It is my hope that you will opt for the most eco-friendly solutions that are available and that ultimately we reach a point where sewage pollution of our groundwater, rivers and streams is a distant memory of a temporary madness. However if this book simply serves as a map to helping you achieve some improvement on a site with difficult soil, or prompts you to plant up a small polluted drain beside your property as a low-tech water filter, then it will have been worthwhile.

While I have tried to be as comprehensive as possible in outlining the different treatment options, be aware that new systems enter the market regularly, and hybrids and adaptations exist, making categorisation difficult. This is particularly the case for challenging sites, where innovation is essential to achieve good results.

While this subject doesn't necessarily make for the most popular dinner conversation ever, it's still better dealt with than not. I wish you every success in your endeavours towards keeping your groundwater or stream a little cleaner for this process.

Glossary

20/30 Effluent: An effluent with a BOD concentration of 20mg/litre and a suspended solids concentration of 30mg/litre.

Activated Sludge: 'Activated Sludge' is a secondary treatment method that aerates sewage effluent to provide the right environment for bacteria and other microorganisms to grow to their maximum capacity in suspension, and thus treat the effluent.

Aquifer: An underground reservoir of water, particularly used in the context of feeding springs, wells and municipal water supplies.

BAF/SAF: Biological/Submerged Aerated Filter – a mechanical treatment system in which air is blown up through a tank of effluent in which a plastic media provides a framework on which bacteria can grow and develop. It is these bacteria that clean up the waste in the oxygen rich environment of the tank.

BOD: Biochemical Oxygen Demand, a measure of the degree of oxygen consumption by microorganisms feeding on the food value within polluted water.

Buffer Zone: Area of ground between a potential pollution source and a watercourse or well-head. Usually these are planted strips between fields and streams or drains, where runoff water can filter through grass or wooded verges before reaching the watercourse, being filtered on its way through the buffer area.

Catchment: A river/stream catchment is the entire area of land that feeds into that river/stream. If a drop of rain falls or a percolation area discharges effluent within that catchment, it will ultimately end up in that river or stream.

Constructed Wetland: In this book, the term constructed wetland is typically used to describe a 'free water', 'surface flow' or 'soil based' treatment wetland.

Coppicing: Cutting back trees, willows in this book, close to ground level during the dormant season on a rotation basis to encourage regrowth of many stems and to keep growth fresh.

Eutrophication: The enrichment of rivers and lakes with nutrients such as nitrates and phosphates, detracting from their health and habitat value.

Faecophobia: The personal or cultural fear or abhorrence of faecal material (as opposed, for example, to the fear of contamination of drinking water sources by sewage pollution *per se*).

Ferric or Alum Dosing: Ferric Sulphate and Aluminium Sulphate are two chemicals that are sometimes added to sewage effluent to remove phosphates. The phosphates bind preferentially to the Ferric and Alum compounds which then settle out in the sludge where they can be removed.

Freeboard Capacity: The height of available space in which a tank or wetland may be raised without overflowing. This is typically factored into the design of soil-based constructed wetlands to lengthen the system lifetime.

Gravel Reed Bed: A lined gravel-filled basin planted with selected wetland plant species for optimum treatment of septic tank or secondary treated effluent.

HF Reed Bed: A Horizontal-Flow gravel reed bed.

Humanure: Just like it sounds – a rich resource of nutrients and biomass that can be composted to return safely to build up soil, but is typically wasted by adding to potable water for disposal to sewer.

MBBR: Moving Bed Bioreactor – a form of BAF/SAF mechanical treatment unit in which the plastic media is free-floating within the oxygenated tank rather than fixed in position.

MBR: Membrane Bioreactor – a mechanical treatment unit similar to an activated sludge plant in which the final effluent undergoes membrane filtration rather than settlement alone to filter out the biosolids generated in the aeration process.

P.E.: Population Equivalent; the number of people contributing to a sewage treatment system, or more specifically, the number of 'people equivalent' in terms of organic nutrient loading or flow volume (assuming that 1pe has a BOD load of 60g/p/d and flow volume of 150 litres/p/d).

Primary or Preliminary Treatment: The initial settlement stage in the wastewater treatment process; typically provided by a septic tank in a domestic context.

RBC: Rotating Biological Contactor – a method of secondary treatment that uses a series of plastic discs to provide a structure for bacteria to adhere to; the rotation of which provides a cycle of submerged and exposed conditions for keeping the bacteria in an aerobic environment.

SBR: Sequencing Batch Reactor – a method of secondary treatment that uses batch loading of a series of treatment chambers as part of its treatment process.

Secondary Treatment: The effluent treatment process of providing sufficient aeration to the effluent in order to allow naturally occurring (or introduced) microorganisms to grow unchecked and thus reduce the organic loading or BOD of the effluent. Suspended solids are settled out of suspension as an integral part of the treatment process.

Source Separation Systems: These are sewage infrastructure that provides separation of urine and/or faecal solids from the flush water, at or close to the toilet; or instead of the toilet. The simplest example is the urinal.

SS: Suspended Solids, literally fine solids in suspension within the water.

Sterilisation: UV, ozone and chlorine are some methods of removing or reducing bacteria numbers in treated effluents.

Surface water discharge: Typically septic tanks discharge to ground via a percolation area. Surface water discharges occur where effluent is piped to a river, lake, stream or drain.

T/P Value: A measure of the percolation value of the soil. This is the time taken (minutes) for water to drop 25mm in a percolation hole, in accordance with the guidance in the EPA *Code of Practice*. The T value is the drop recorded close to the actual percolation trench base, the P value is the same test carried out close to ground surface to check the topsoil percolation rate.

Tertiary Treatment: Additional filtration of secondary treated effluent, typically used in the context of reducing concentrations of nitrates and phosphates.

VF Reed Bed: A top loaded Vertical-Flow gravel reed bed system.

Zero Discharge Willow Facility: A willow planted basin designed to receive septic tank effluent and evaporate 100% to air. Thus there is no discharge to the receiving environment, either to ground or to surface waters.

APPENDIX 2

References and Resources

Cooper PF, GD Job, MB Green and RBE Shutes (1996) *Reed beds and constructed wetlands for wastewater treatment*. WRc Swindon, Wiltshire.

Costello C (ed.) (1993) *Proceedings of conference on constructed wetlands for waste water treatment in Ireland*. Maxpro, Kinsale.

Department of the Environment, Heritage and Local Government (2010) *Integrated constructed wetland guidance document for farmyard soiled water and domestic wastewater applications*. DEHLG, Dublin.

Dubber D and L Gill (2013) *STRIVE Report Series No. 108: Water saving technologies to reduce water consumption and wastewater production in Irish households*. Environmental Protection Agency, Wexford.

Fitter R, A Fitter and A Farrer (1984) *Collins Pocket Guide: Grasses, Sedges, Rushes and Ferns of Britain and Northern Europe*. HarperCollins, London.

Fitter R and R Manuel (1994) *Collins Photo Guide: Lakes, Rivers, Streams and Ponds of Britain and Northwest Europe*. HarperCollins, London.

Fitter R, A Fitter and M Blamey (1996) *Collins Pocket Guide to Wild Flowers of Britain and Northwest Europe*. HarperCollins, London.

Hammer D (ed.) (1989) *Constructed Wetlands for Wastewater Treatment. Municipal, Industrial, Agricultural*. Lewis Publishers Inc., Chelsea, MI, USA.

Hammer DA (1993) Constructed wetlands for wastewater treatment, an overview of a low-cost technology. In: Costello C (ed.) *Proceedings of conference on constructed wetlands for wastewater treatment in Ireland*. Maxpro, Kinsale, Co. Cork.

Harty F (1999) *Constructed wetland monitoring programme at Ballymaloe House, Midleton, Co. Cork, and health of Rooska Stream, the receiving water body*. Poster presentation at Environmental Researchers Colloquium 1999, Johnstown Castle, Wexford.

Harty F and ML Otte (2003) Constructed wetlands for treatment of wastewater. In: Otte ML. (ed.) *Wetlands of Ireland: Distribution, ecology, uses and economic value*. UCD Press, Dublin.

Haycock NE, TP Burt, KWT Goulding and G Pinay (1996) *Buffer zones: Their processes and potential in water protection*. Quest Environmental, Hertfordshire.

Hutchinson L (2005) *Ecological Aquaculture: A Sustainable Solution*. Permanent Publications, Hampshire.

Jenkins J (2005) *The Humanure Handbook: A Guide to Composting Human Manure*. Jenkins Publishing, Pennsylvania, USA.

Kadlec RH and SD Wallace (2009) *Treatment Eetlands*, second edition. CRC Press, Florida, USA.

Kiely G (1997) *Environmental Engineering*. McGraw Hill, London.

Martin WK (1991) *The New Concise British Flora*. Bloomsbury Books, London.

Meyer K (1989) *How to Shit in the Woods – An Environmentally Sound Approach to a Lost Art*. Ten Speed Press, Berkley, Ca, USA.

Mulqueen J, M Rogers, B Gallagher, E Waldron and B Fehily (1998) *Small scale wastewater treatment systems, literature review*. R&D Report Series No.3, EPA, Wexford.

Mulqueen J, M Rodgers, G O'Leary and G Carty (2000) *EPA Wastewater Treatment Manuals: Treatment Systems for Single Houses*. EPA, Wexford.

Rogers M, J Mulqueen, G Carty and G O'Leary (1999) *EPA Wastewater Treatment Manuals: Treatment Systems for Small Communities, Businesses, Leisure Centres and Hotels*. EPA, Wexford.

Sale C (1929) *The Specialist*. Putnam and Co., London.

Seymour J (2003) *The New Complete Book of Self Sufficiency: The Classic Guide for Realists and Dreamers*. Dorling Kindersley, London.

Steinfeld C (2004) *Liquid Gold: The Lore and Logic of Using Urine to Grow Plants*. Green Books, Devon.

Waste management (use of sewage sludge in agriculture) regulations S.I. No. 148 of 1998 and amendment S.I. No. 267 of 2001.

Whitefield P (2004) *The Earth Care Manual: A Permaculture Handbook for Britain and Other Temperate Climates*. Permanent Publications, Hampshire.

Winblad U and M Simpson-Hébert (eds.) *Ecological Sanitation*, revised and enlarged edition. SEI, Stockholm, Sweden, 2004.

Woods-Ballard B, R Kellagher, P Martin, C Jeffries, R Bray and P Shaffer (2007) *The SUDS Manual*. CIRIA, London.

Useful Internet References

This is a very condensed list of Internet references that I have found useful in designing and guiding clients with wastewater treatment systems. I haven't included secondary treatment system suppliers, wetland and reed bed designers, dry toilet suppliers etc. because a quick Internet search will usually yield an up to date list. Please bear in mind that website addresses are updated from time to time, so if any of the following links are broken, just do a search on the subject matter and you'll probably find what you're looking for.

Building Regulations, part H (2010):
www.environ.ie/en/Publications/DevelopmentandHousing/BuildingStandards/
FileDownLoad,24906,en.pdf

Centre for Alternative Technology's Small-Scale Sewage Treatment and Composting Toilets information sheet:
http://info.cat.org.uk/sites/default/files/documents/SewageTreatmentAndCompostToilets.pdf

Department of the Environment, Community and Local Government's grant scheme for septic tank upgrades can be found at:
www.epa.ie/water/wastewater/info/grants/#.U0e9Dygww7s

Department of the Environment and Local Government / Environmental Protection Agency / Geological Service of Ireland, Groundwater Protection Response document can be found here:
www.gsi.ie/NR/rdonlyres/20AF2A5A-8DCC-45B3-96C8-5AC8CBA86B83/0/sing_hse.pdf

Dublin City Council's grey water and rainwater harvesting tips:
www.dublincity.ie/main-menu-services-water-waste-and-environment-your-drinking-water-rainwater-harvesting/grey-water

EPA *Code of Practice*:
www.epa.ie/pubs/advice/water/wastewater/code%20of%20practice%20for%20single%20houses/# to download the EPA *Code of Practice* for Wastewater Treatment Systems for Single Houses.

EPA Domestic Wastewater Treatment System Inspection form:
http://epa.ie/pubs/advice/water/wastewater/DWWTS_Inspections_Form.pdf

Ecological Sanitation Research website is an invaluable resource for dry toilet systems, with Ecological Sanitation a good concise document if you want to broaden your horizons:
www.ecosanres.org/pdf_files/Ecological_Sanitation.pdf

Integrated Constructed Wetland Guidance Document:
www.environ.ie/en/Publications/Environment/Water/FileDownLoad,24931,en.pdf

Joseph Jenkins' Humanure Handbook site. The classic guide to humanure composting:
http://humanurehandbook.com

Oasis Design grey water information. An excellent resource for ecologically friendly grey water treatment and disposal:
www.oasisdesign.net

UK Environment Agency's grey water guide: *Greywater for domestic users: an information guide*, May 2011, Environment Agency, Bristol, UK:
www.sswm.info/sites/default/files/reference_attachments/ENVIRONMENT%20 AGENCY%202011%20Greywater%20for%20Domestic%20Users.pdf

Septic Tank Options and Alternatives was prepared using open source and freeware programs. These were Open Office Writer (www.openoffice.org/download/); drawings were completed in Draftsight (www.3ds.com/products-services/draftsight/overview/); graphs were prepared in yEd Graph Editor (www.yworks.com/en/products_yed_about.html).

For other websites around sewage treatment and sustainable construction, check the FH Wetland Systems links page here: www.wetlandsystems.ie/links.html

Homeowner Septic Tank Inspection Record Sheet

This record sheet may be used to keep tabs on your own inspections over time, so photocopy this sheet and keep a copy each time you inspect your system yourself. This record sheet is based on the EPA *Code of Practice* recommendations, and may be helpful for a formal septic tank inspection process, but does not replace it.

Tick sheet for septic tank self-inspection and recording:

Homeowner name		
Site address		
Date		
Background information	Is your well or neighbouring wells, if present, free of contamination by faecal bacteria to the best of your knowledge?	
	Does your toilet flush freely?	
Distribution device	Is the distribution box present between your septic tank and percolation area?	
	Is it intact and providing an even flow to each outlet pipe?	
Percolation area	Is the percolation area clearly identifiable?	
	Is it free of surface ponding?	
	Are the vent pipes present and intact?	
	Are vent pipes dry inside and free of obstruction?	
	Is the ground free of surface damage by vehicular activity, heavy animals, sports or other activities?	
	Are adjacent drains and streams free from signs of sewage fungus, algal growth and black anaerobic sludges?	

Septic tank	Sludge depth prior to desludging (cm):	
	Is the sludge depth ≤30cm?	
	Scum layer thickness (cm):	
	Is the scum layer ≥10cm from outlet pipe level?	
	Is the inlet T-piece pipe free flowing?	
	Is the outlet T-piece pipe free flowing (including outlet screen/filter where present; and is the outlet screen intact and functioning effectively)?	
	Desludging date:	
	Contractor name:	
	Do you have a copy of your certificate of service?	
	Is the baffle wall in place and fully intact, dividing section 1 and 2?	
	Is the baffle wall clear of debris?	
	Is the tank free from signs of leakage in or out?	
	Is the sewer network free from signs of water ingress?	

APPENDIX 5

Treatment System Evaluation Sheet

This evaluation sheet is reproduced from Appendix C of the EPA (2000) *Wastewater Treatment Manuals: Treatment Systems for Single Houses* (Updated from £ to €).

Factor	Treatment Option No.1	Treatment Option No.2
Capital cost €		
Construction cost prior to delivery €		
Additional costs prior to commissioning €		
Annual running cost €/annum		
Installation and commissioning service available		
Maintenance agreement available		
Cost of annual maintenance agreement €		
Design criteria*		
Performance - % reduction in BOD, COD, TSS		
Performance - % reduction in Total P and Total N		
Performance - % reduction in Faecal *coliforms*		
Beneficial uses of the receiving water		
Guarantee available		
Agrément certification		
Recommendations from other users		
Expected life of the system		
Power requirements kW/d		
Power requirements – single phase/three phase		
Ease of operation		

Daily, weekly and annual maintenance requirements		
Licence required (Water Pollution Act Licence)		
Access requirements for sludge removal		
Sludge storage capacity (m³)		

* In the case of biofilm systems the organic and hydraulic loading rates in $g/m^2/d$ and $l/m^2/d$ respectively should be quoted.

APPENDIX 6

EPA Interpretation of Percolation Test Results

Table 6.3 Interpretation Of Percolation Test Results, from the EPA *Code of Practice* (2009)

Percolation test results	Interpretation
T > 90	Site is unsuitable for development of any on-site domestic wastewater treatment system discharging to ground. Site may be deemed suitable for treatment system discharging to surface water in accordance with Water Pollution Act licence.
T < 3	Retention time in the subsoil is too fast to provide satisfactory treatment. Site is unsuitable for secondary-treated on-site domestic wastewater systems. However, if effluent is pretreated to tertiary quality then the site will be hydraulically suitable to assimilate this hydraulic load. P-test should be undertaken to determine whether the site is suitable for a secondary treatment system with a polishing filter at ground surface or overground. Sites may be deemed suitable for discharge to surface water in accordance with Water Pollution Act licence.[1]
3 ≤ T ≤ 50	Site is suitable for the development of a septic tank system or a secondary treatment system discharging to groundwater.
50 < T < 75	Wastewater from a septic tank system is likely to cause ponding at the surface of the percolation area. Not suitable for a septic tank system. May be suitable for a secondary treatment system with a polishing filter at the depth of the T-test hole.
70 ≤ T ≤ 90	Wastewater from a septic tank system is likely to cause ponding at the surface of the percolation area. Not suitable for a septic tank system. Site unsuitable for polishing filter at the depth of the T-test hole. P-test should be undertaken to determine whether the site is suitable for a secondary treatment system with polishing filter, i.e. 3 ≤ P ≤ 75, at ground surface or overground.
P < 3	Retention time in the topsoil/subsoil insufficient to provide satisfactory treatment. However, if effluent is pretreated to tertiary state then the site will be hydraulically suitable to assimilate the hydraulic load. Imported suitable material may be deemed acceptable as part of site improvement works.

$3 \leq P \leq 75$	Site is suitable for a secondary treatment system with polishing filter at ground surface or overground. If the subsoil is classified as CLAY, carry out a particle size distribution and refer to I.S. CEN/TR 12566-2:2005.
T not possible due to high water table	

1   Most local authorities do not grant water pollution discharge licences to single dwellings and the site assessor is advised to contact the Environment Section for advice.

APPENDIX 7.1 Pollution Pathways Showing Movement of Effluent from Use to Receiving Environment

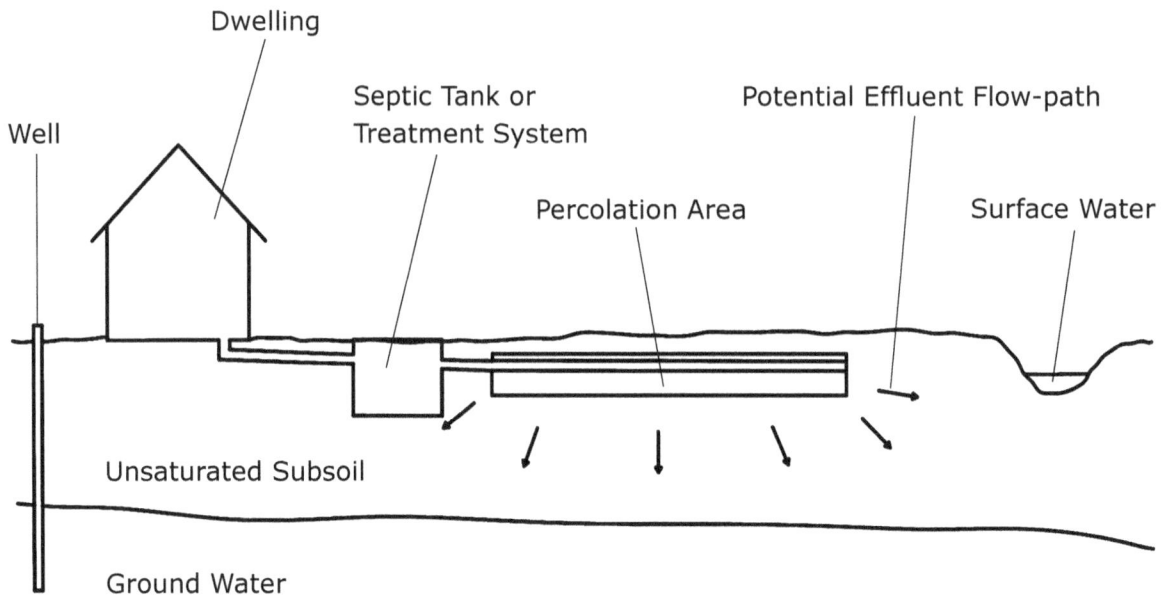

APPENDIX 7.2 A Typical Septic Tank

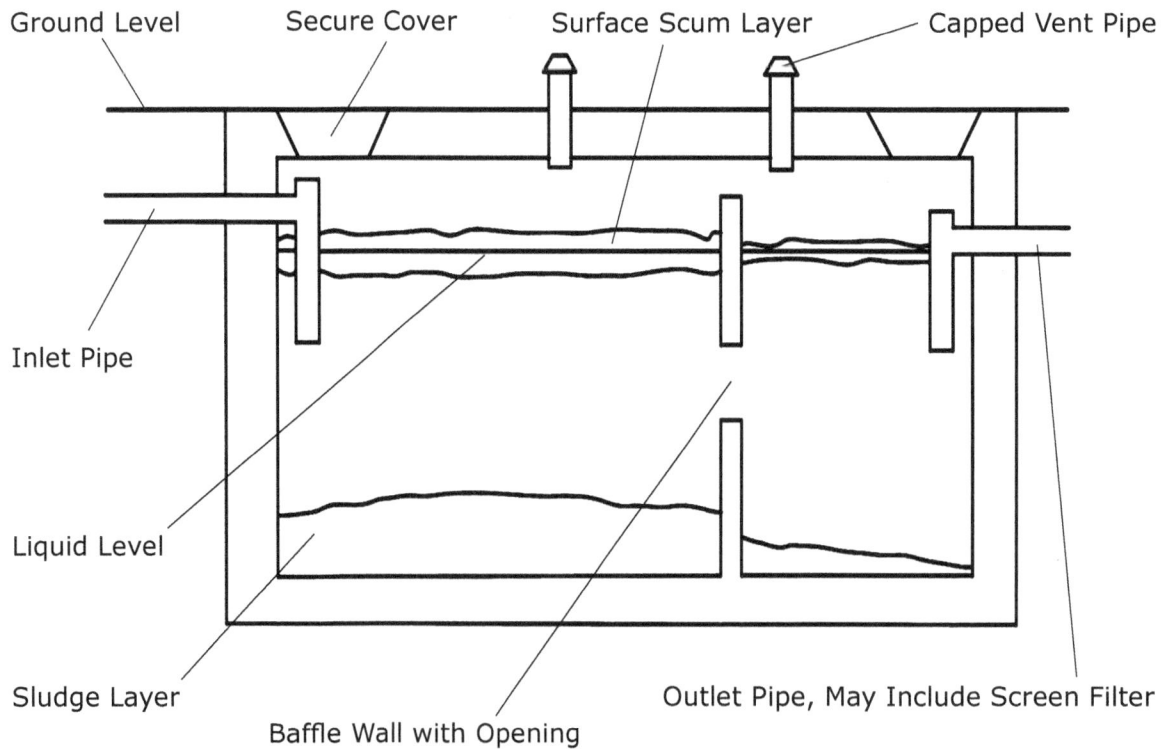

APPENDIX 7.3   Soil-based Constructed Wetland System

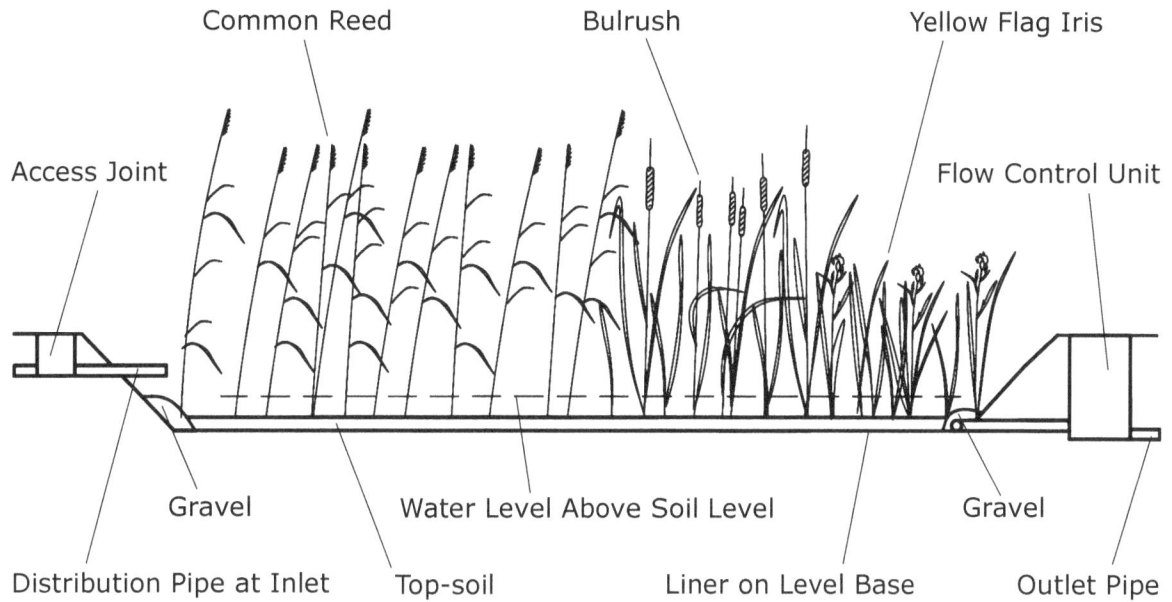

Common Reed — Bulrush — Yellow Flag Iris — Flow Control Unit — Access Joint — Gravel — Water Level Above Soil Level — Gravel — Distribution Pipe at Inlet — Top-soil — Liner on Level Base — Outlet Pipe

APPENDIX 7.4   Horizontal-flow Gravel Reed Bed System

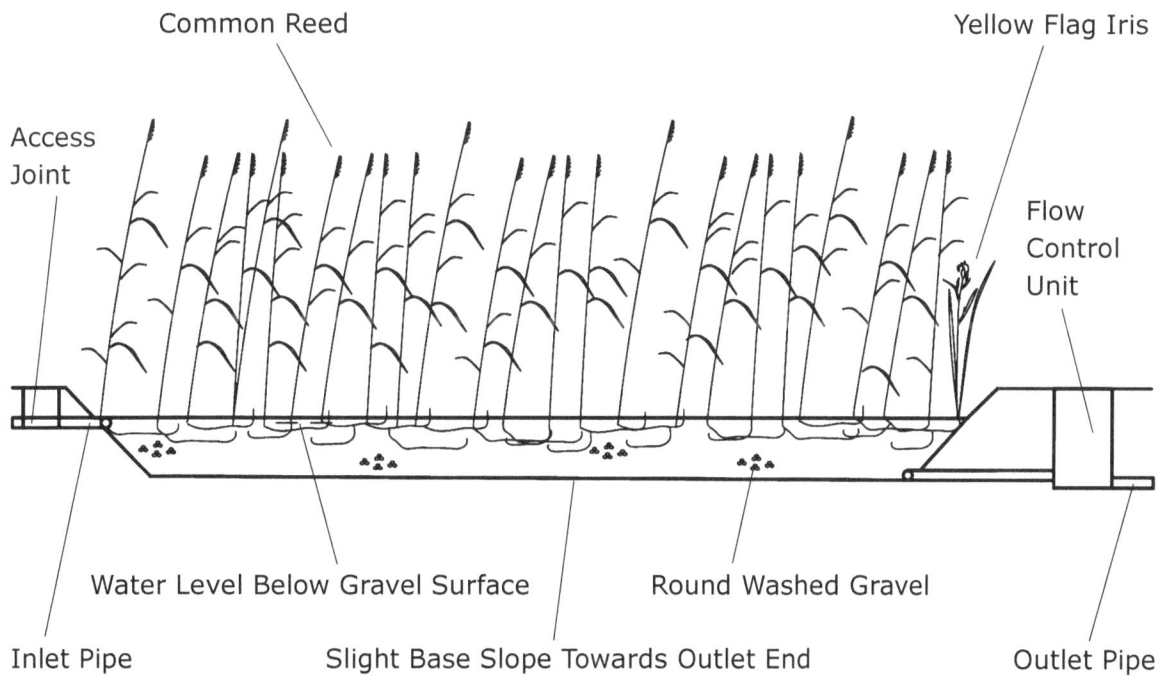

Common Reed — Yellow Flag Iris — Access Joint — Flow Control Unit — Water Level Below Gravel Surface — Round Washed Gravel — Inlet Pipe — Slight Base Slope Towards Outlet End — Outlet Pipe

APPENDIX 7.5   **Vertical-flow Gravel Reed Bed System**

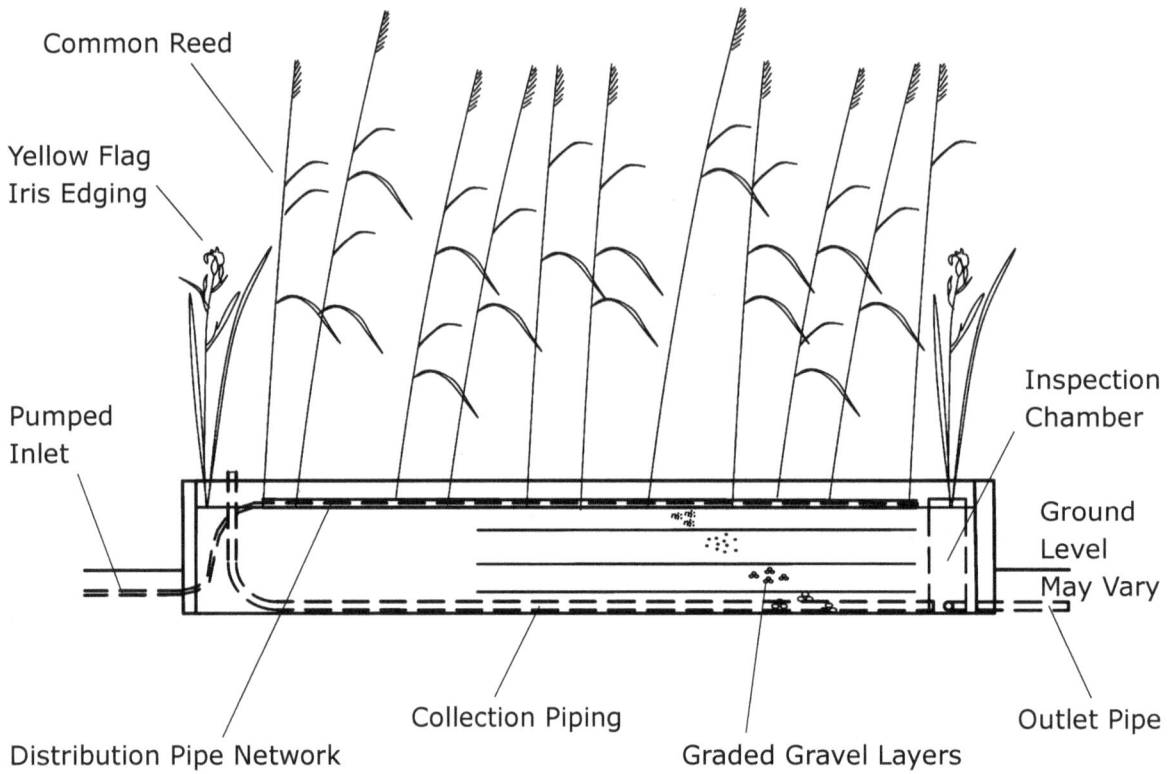

Common Reed

Yellow Flag
Iris Edging

Pumped
Inlet

Inspection
Chamber

Ground
Level
May Vary

Distribution Pipe Network

Collection Piping

Graded Gravel Layers

Outlet Pipe

APPENDIX 7.6   **Zero Discharge Willow Facility**
(Details reproduced with permission from
the Centre for Recirkulering, Denmark)

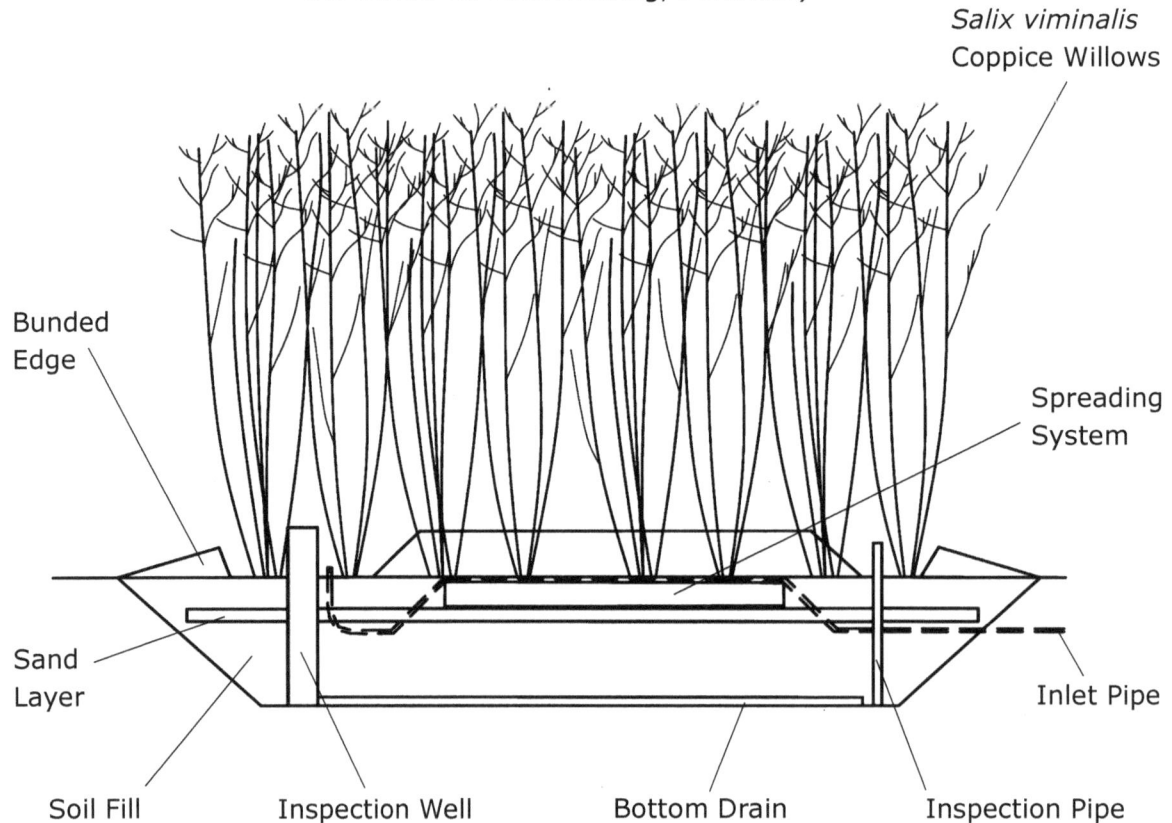

Salix viminalis
Coppice Willows

Bunded
Edge

Spreading
System

Sand
Layer

Inlet Pipe

Soil Fill

Inspection Well

Bottom Drain

Inspection Pipe

APPENDIX 8.1   Septic Tank and Percolation Area

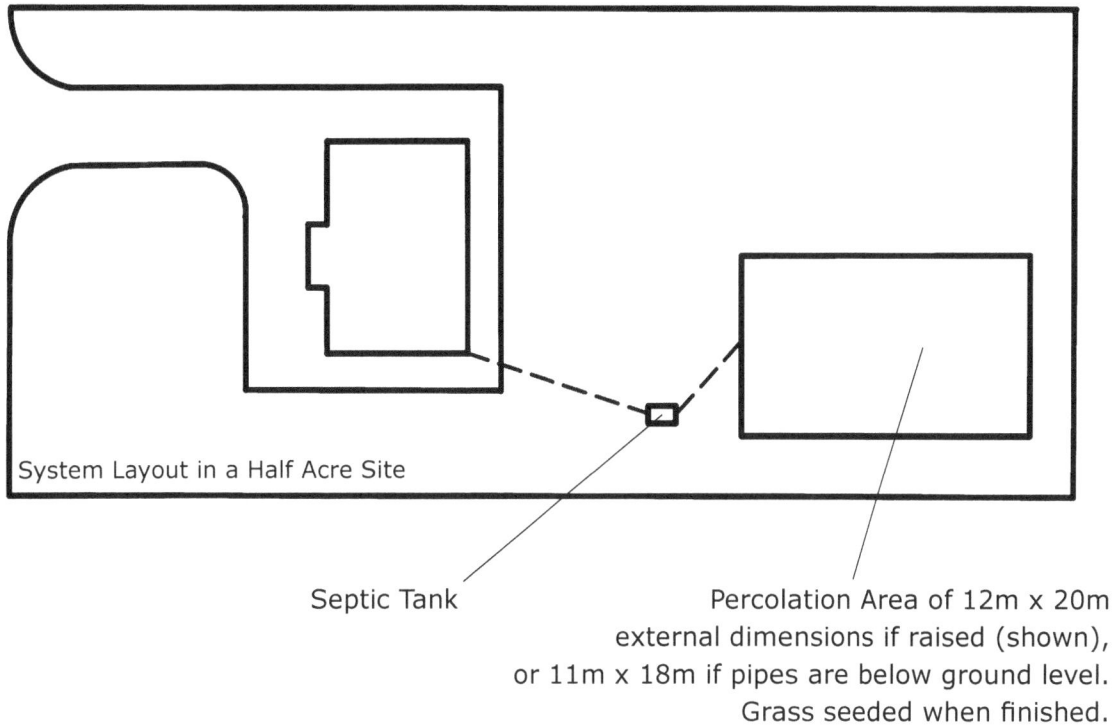

System Layout in a Half Acre Site

Septic Tank

Percolation Area of 12m x 20m
external dimensions if raised (shown),
or 11m x 18m if pipes are below ground level.
Grass seeded when finished.

APPENDIX 8.2   Mechanical or Packaged Filter Unit and Soil Polishing Filter

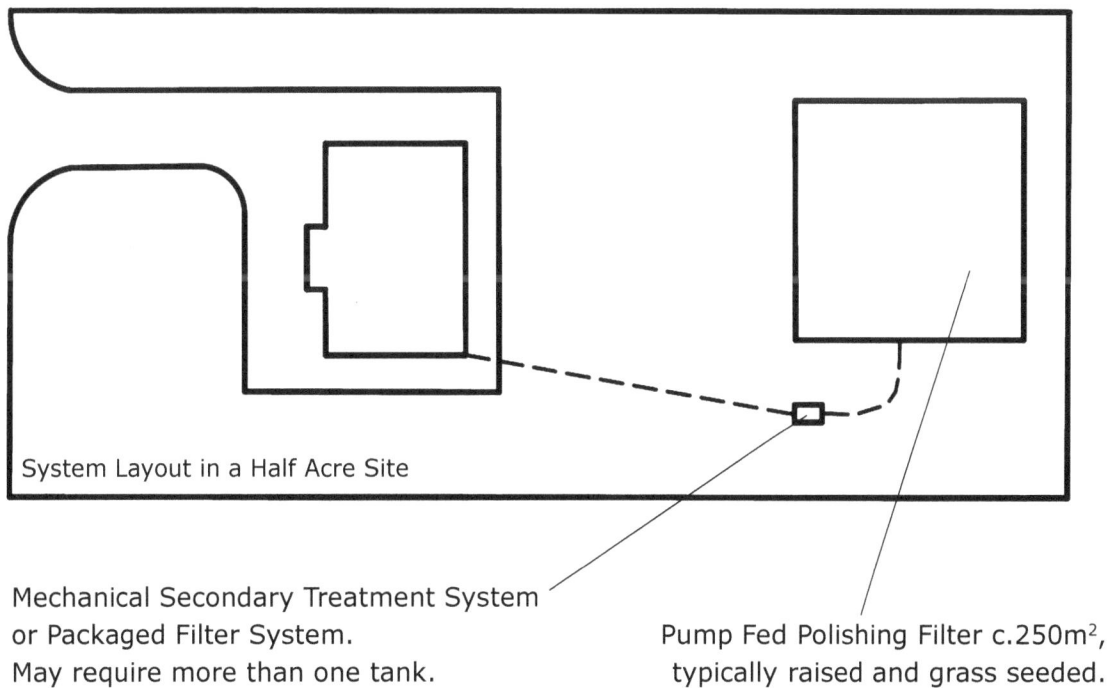

System Layout in a Half Acre Site

Mechanical Secondary Treatment System
or Packaged Filter System.
May require more than one tank.

Pump Fed Polishing Filter c.250m^2,
typically raised and grass seeded.

APPENDIX 8.3   Septic Tank, Constructed Wetland and Soil Polishing Filter

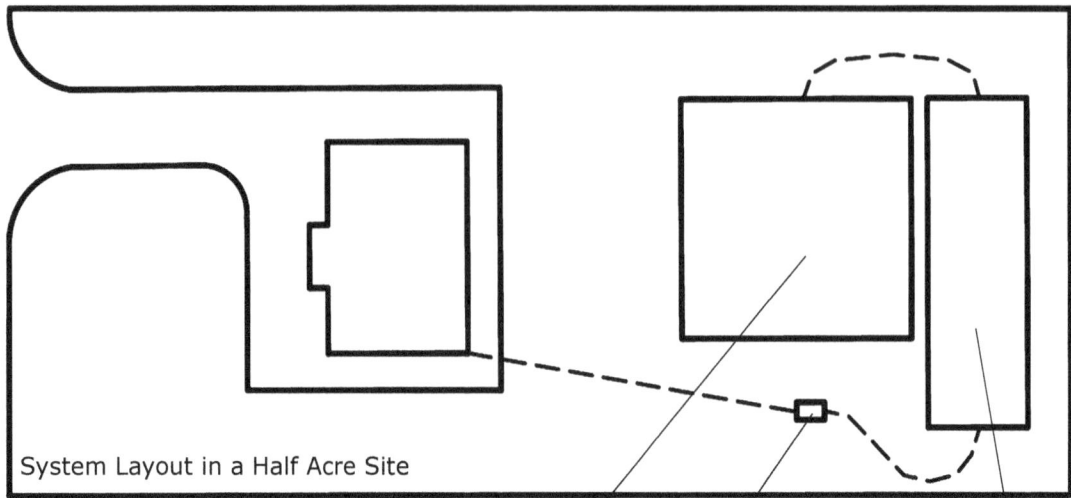

System Layout in a Half Acre Site

Pump Discharge Polishing Filter of c.250m^2, typically raised and grass seeded.

Septic Tank

Soil-Based Constructed Wetland System 7m x 22m

APPENDIX 8.4   Two Septic Tanks In Series, Horizontal-flow Reed Bed and Polishing Filter

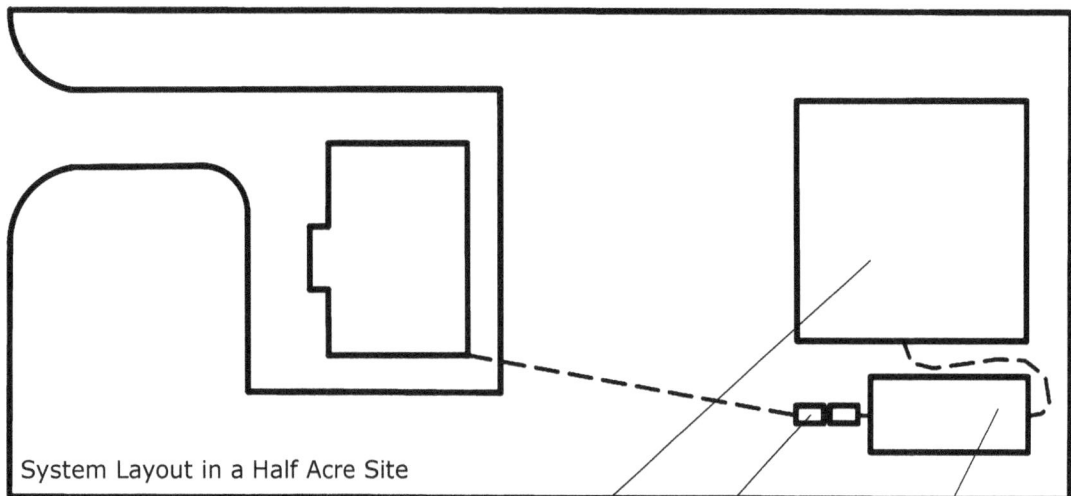

System Layout in a Half Acre Site

Pump Discharge Polishing Filter 250m^2, typically raised and grass seeded.

Gravel Reed Bed 5m x 11m

2 x Twin Chamber Septic Tanks

APPENDIX 8.5   Mechanical or Packaged Filter Unit, Constructed Wetland for Tertiary Treatment and Distribution Area

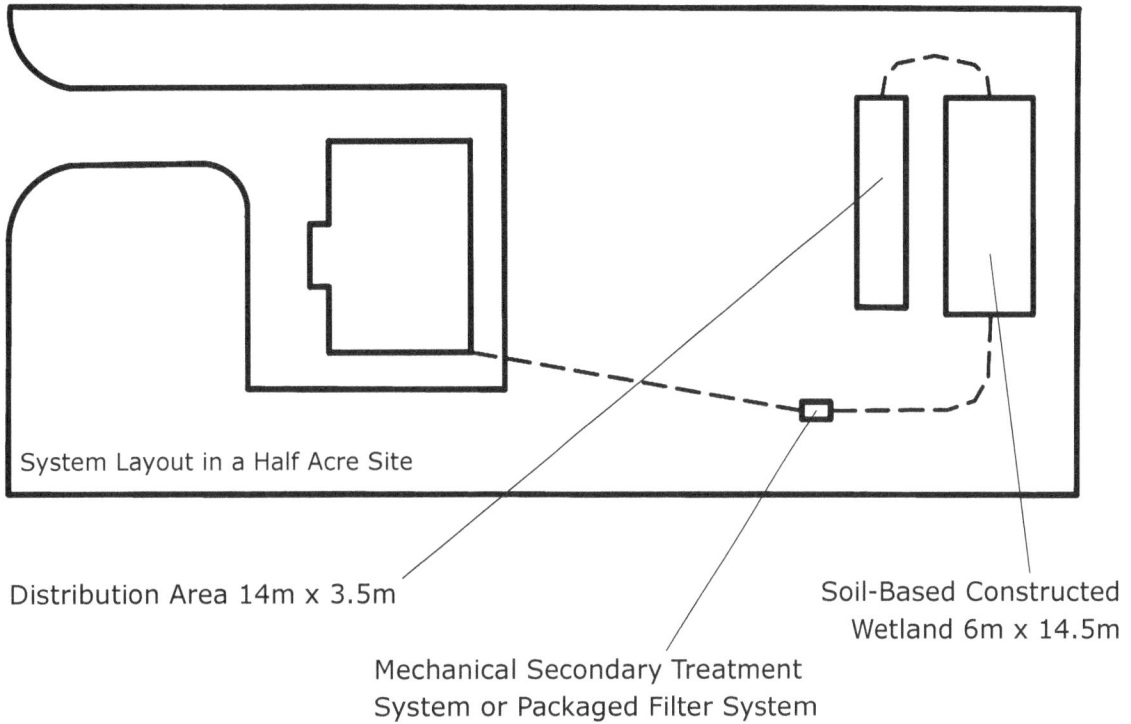

System Layout in a Half Acre Site

Distribution Area 14m x 3.5m

Mechanical Secondary Treatment System or Packaged Filter System

Soil-Based Constructed Wetland 6m x 14.5m

APPENDIX 8.6   Septic Tank, Secondary and Tertiary Treatment Constructed Wetlands and Hybrid Percolation System

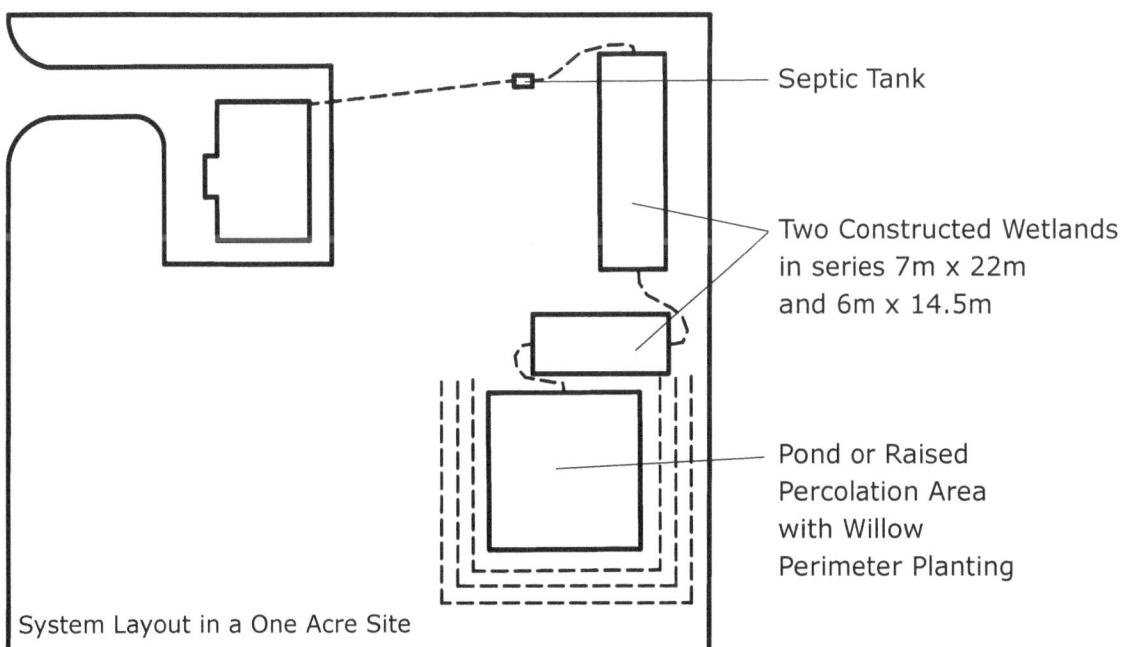

Septic Tank

Two Constructed Wetlands in series 7m x 22m and 6m x 14.5m

Pond or Raised Percolation Area with Willow Perimeter Planting

System Layout in a One Acre Site

APPENDIX 8.7 Two Septic Tanks in Series, Horizontal-flow Reed Bed, Vertical-flow Reed Bed, Hybrid Willow Percolation System

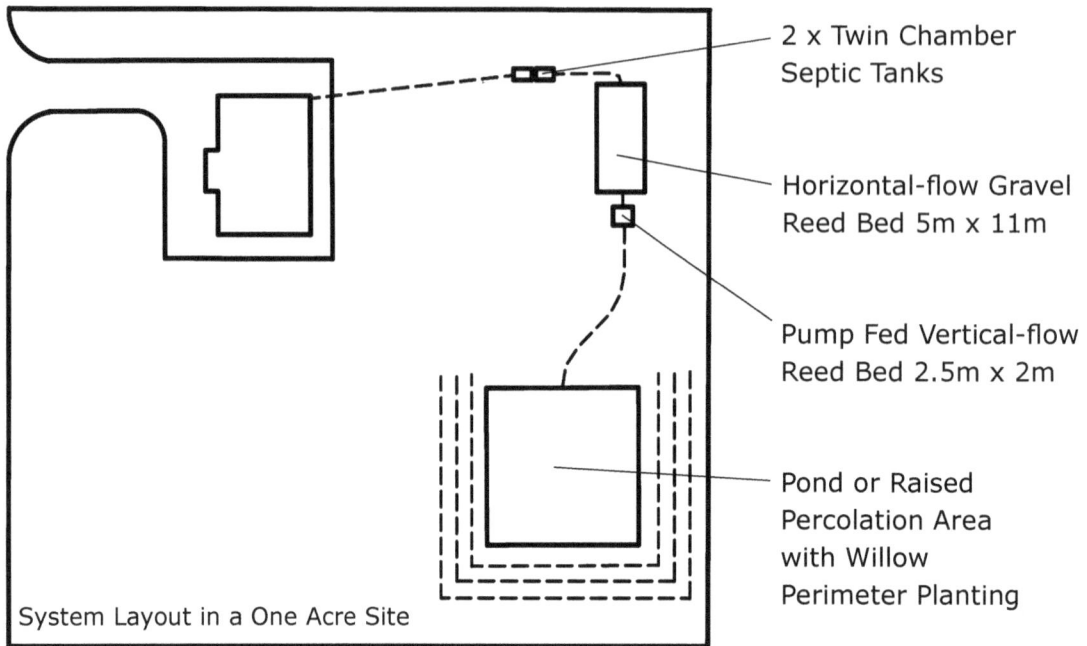

2 x Twin Chamber Septic Tanks

Horizontal-flow Gravel Reed Bed 5m x 11m

Pump Fed Vertical-flow Reed Bed 2.5m x 2m

Pond or Raised Percolation Area with Willow Perimeter Planting

System Layout in a One Acre Site

APPENDIX 8.8 Two Septic Tanks In Series, Vertical-flow Reed Bed, Horizontal-flow Reed Bed and Distribution Area

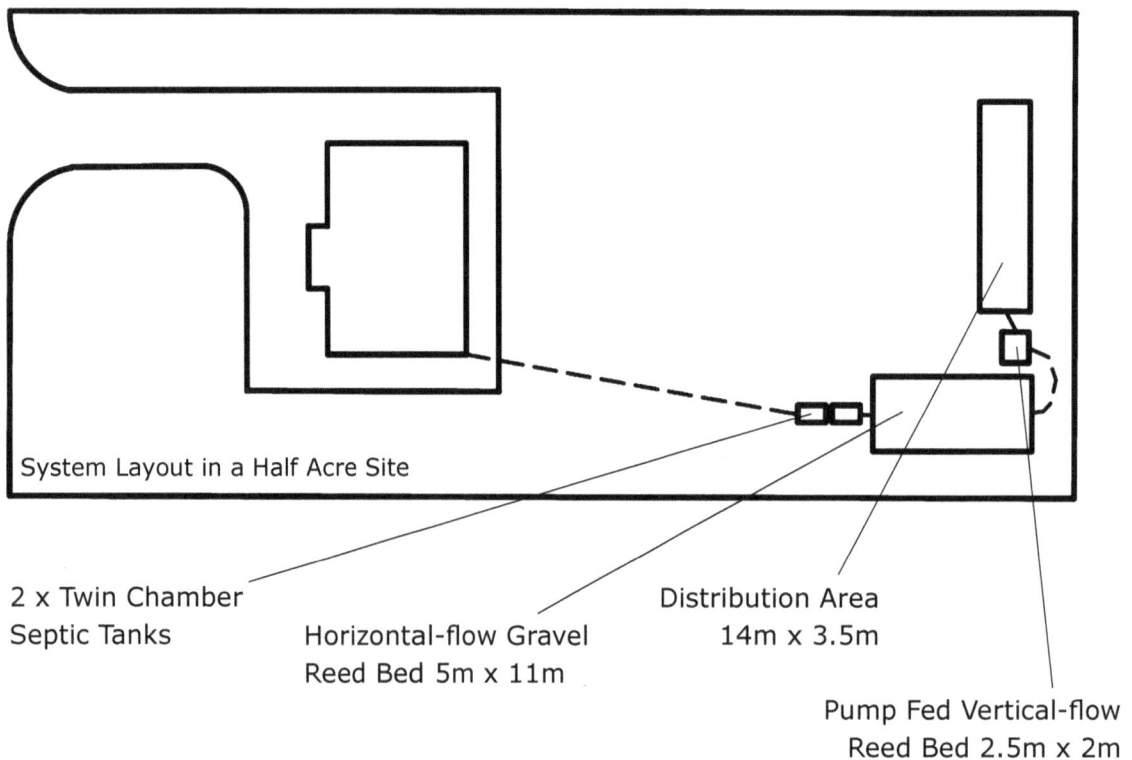

System Layout in a Half Acre Site

2 x Twin Chamber Septic Tanks

Horizontal-flow Gravel Reed Bed 5m x 11m

Distribution Area 14m x 3.5m

Pump Fed Vertical-flow Reed Bed 2.5m x 2m

APPENDIX 8.9 Mechanical or Packaged Filter Unit, Vertical-flow Reed Bed and Distribution Area

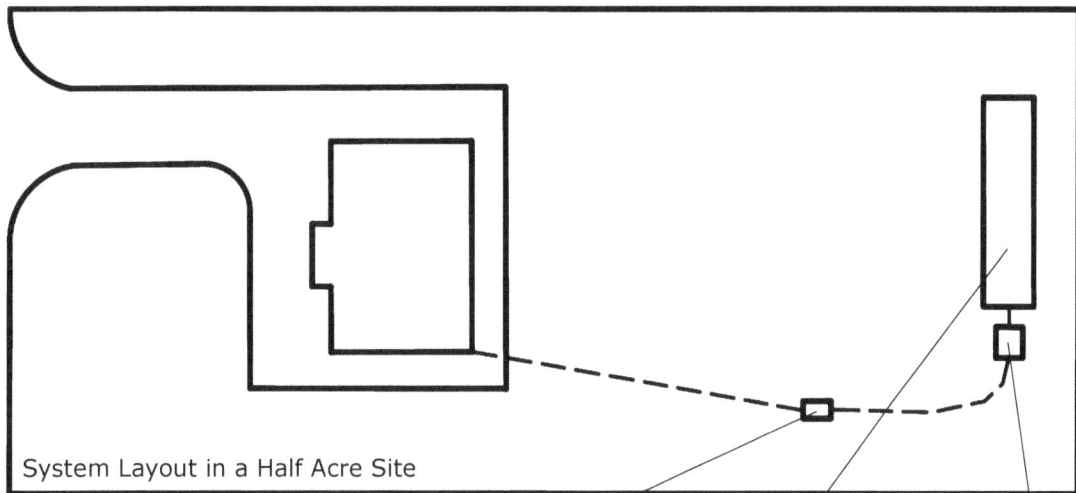

System Layout in a Half Acre Site

Mechanical Secondary Treatment
or Packaged Filter System.
May require more than one tank.

Distribution Area
c.14m x 3.5m

Horizontal- or
Vertical-flow Reed
Bed 2.5m x 2m

APPENDIX 8.10 Septic Tank and Willow Plantation

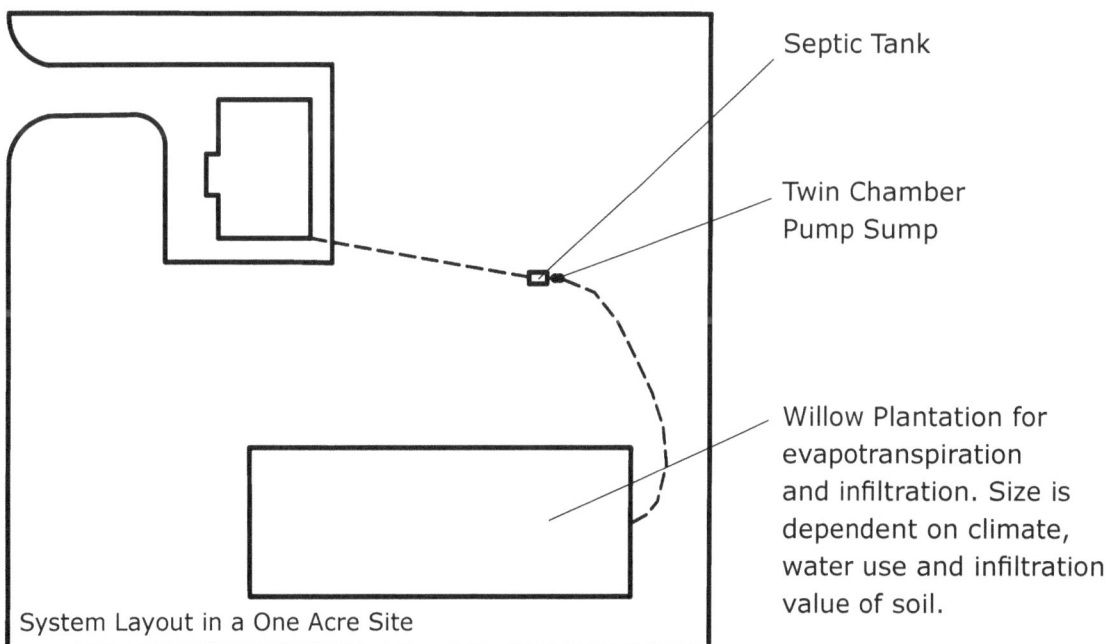

Septic Tank

Twin Chamber
Pump Sump

Willow Plantation for
evapotranspiration
and infiltration. Size is
dependent on climate,
water use and infiltration
value of soil.

System Layout in a One Acre Site

APPENDIX 8.11 **Septic Tank and Twin Basin Zero Discharge Willow Facility**

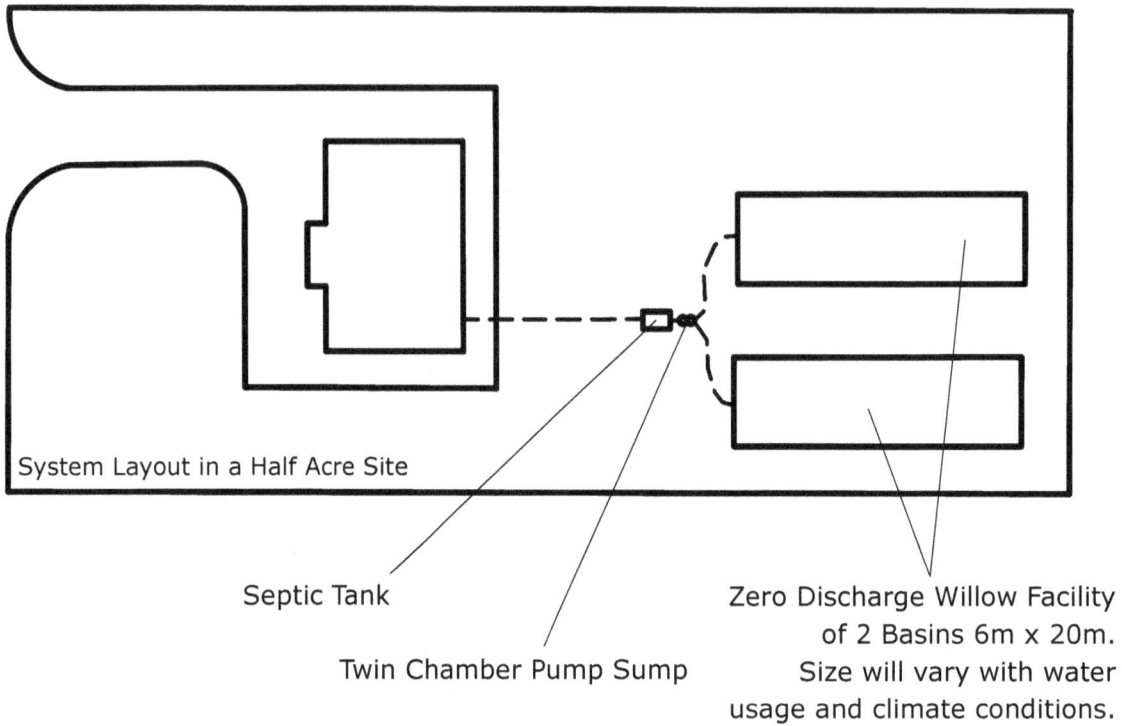

System Layout in a Half Acre Site

Septic Tank

Twin Chamber Pump Sump

Zero Discharge Willow Facility of 2 Basins 6m x 20m. Size will vary with water usage and climate conditions.

APPENDIX 8.12 **Septic Tank and Zero Discharge Willow Facility**

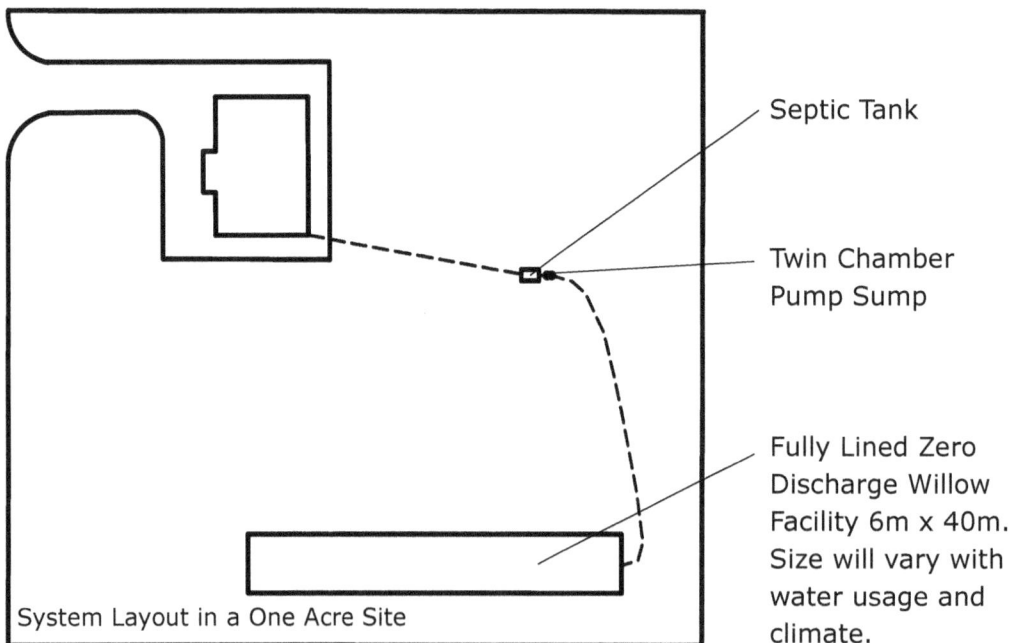

Septic Tank

Twin Chamber Pump Sump

Fully Lined Zero Discharge Willow Facility 6m x 40m. Size will vary with water usage and climate.

System Layout in a One Acre Site

APPENDIX 9   FH Wetland Systems Treatment System Selection Cards

These 'cards' form part of FH Wetland Systems training seminars and workshops for domestic sewage treatment system selection.
The context is a for sewage treatment from a typical 3-bedroom domestic dwelling; 5pe. (persons equivalent). For reed bed and sewage treatment option books, training, seminars, and consultancy contact reeds@wetlandsystems.ie or visit www.wetlandsystems.ie

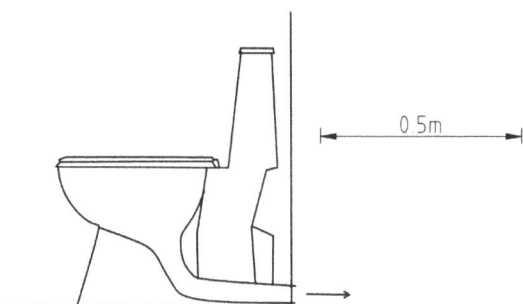

Standard flush toilet
Because you've got to start somewhere in the decision making process.
Note on standard nomenclature:
Black water – from standard flush toilet only.
Grey water – from washing machines, dish washers, sinks, shower, bath.
Stormwater – rainwater runoff from roof and paved surface (not connected to foul sewers).
Also in Australian codes:
Yellow water: Urine and micro flush from separator toilets.
Brown water: Faecal/paper/flush water from separator toilet.
Cost: €200-500 + installation

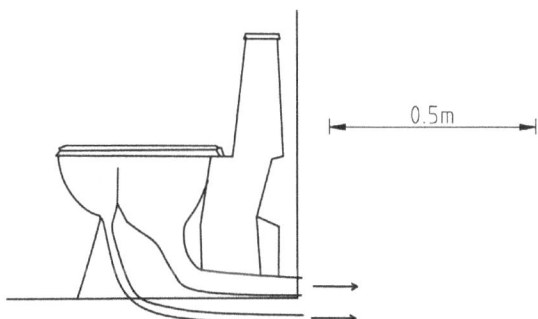

Diverter: Urine diversion toilet (Dubbletten or Wostman)
Removes c 80% of urine at source.
Cost: €900-1300 + installation

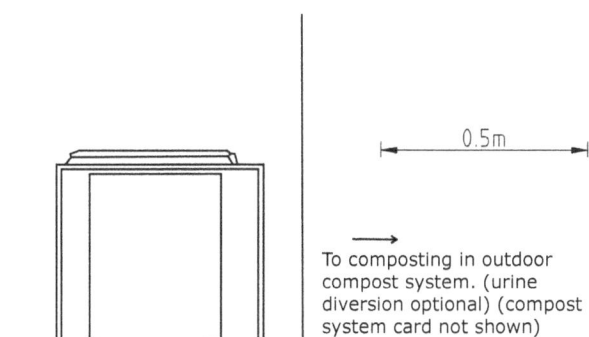

Dry toilet (indoor bucket type shown here)
No water needed.
Many different types and designs available.
May be indoor/outdoor; urine diverting or combined.

To composting in outdoor compost system. (urine diversion optional) (compost system card not shown)

Cost: €0-800 for bucket types and up to 5000 for others.

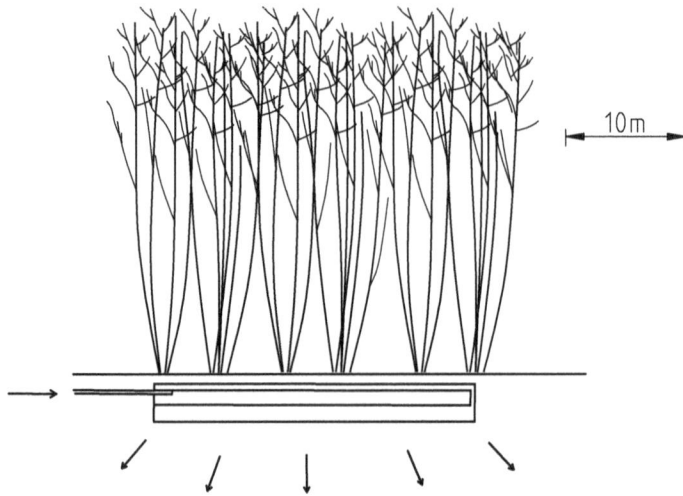

Infiltration: Willow (or comfrey) planted percolation area

Size options:
After settlement: 18m x 1 10m or larger for full N and P removal.
After secondary treatment: 37.5 - 250m^2 (Table 10.1 EPA Code).
After tertiary treatment: 18. 75 - 125m^2 (Table 10.1 EPA Code).
Cost: €1500-5000 installed

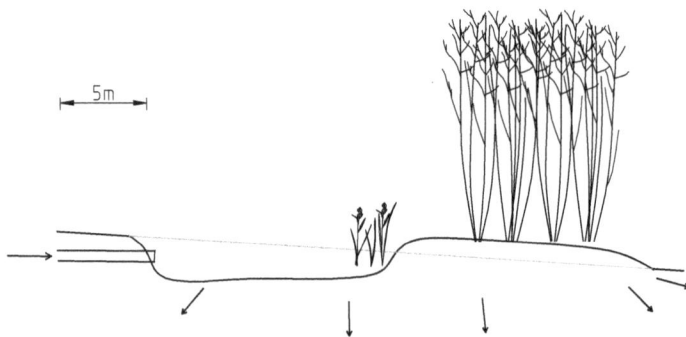

Infiltration: Percolating marsh/pond with willow embankment

Size options:
After settlement: 5m x 20m marsh/pond plus 4m x 25m embankment.
After secondary treatment: 5m x 10m marsh/pond plus 4m x 15m embankment.
After tertiary treatment: 3m x 5m marsh/pond plus 3m x 7m embankment.
Not EPA Code approved; for use on legacy sites with poor percolation.
Cost: €500-2000

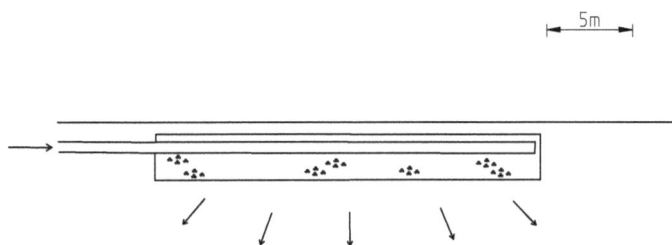

Infiltration: Percolation area (grass cover)

Size options:
After settlement: 18m x 10m or larger for full N and P removal.
After secondary treatment: 37.5 - 250m^2 (Table 10.1 EPA Code).
After tertiary treatment: 18. 75 - 125m^2 (Table 10.1 EPA Code).
Cost: €2000-12000 installed

Treatment: Constructed Wetland (Soil Based)

Size options:
After settlement: 5m x 20m
After secondary treatment:
5m x 10m
Cost: €500-3500

See Appendix 7.3 p.127

Treatment: Horizontal Flow Gravel Reed Bed

After settlement: 5m x 10m min.
After secondary treatment:
1m x 5m min.
Cost: €2500-3500

See Appendix 7.4 p.127

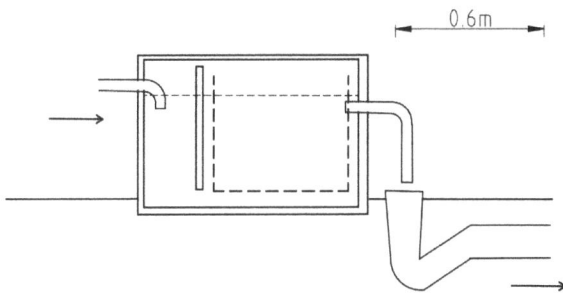

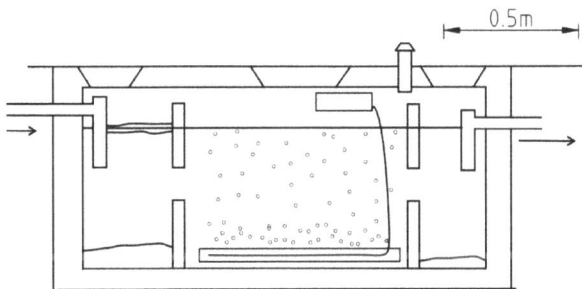

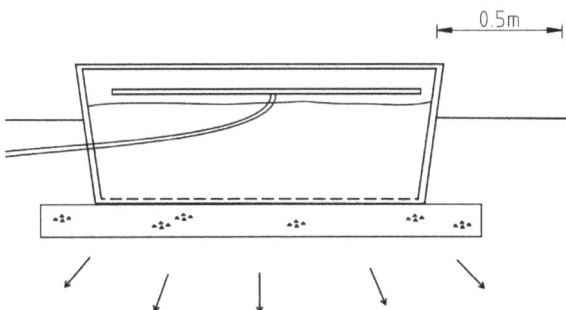

Treatment: Vertical Flow Gravel Reed Bed

After settlement: 4m x 5m
After secondary treatment:
2m x 2.5m
Cost: €2500-3500

See Appendix 7.5 p.128

Disposal to Air: Zero Discharge Willow System

Size options:
Dry county: 6m x 40m
Wet county: 6m x 70m
Pump fed from septic tank.
Cost: €20-50,000

See Appendix 7.6 p.128

Settlement: Grease Trap (grey water only)

c. 0.7m x 0.4m positioned above ground outside kitchen sink or washing machine wall or 4" version in the grey water sewer line.
Cost: €300 + installation

Settlement and treatment combined: Mechanical Treatment System

c. 3m x 2m; Varies with supplier.
Needs electricity.
Many different design types.
Cost: €3500-8000

Treatment: Media Filter System (typically coconut fibre, may also be woodchip)

c. 3m x 2m x 2no. units; Request EPA approved size from supplier.
Typically needs electricity.
Woodchip filter not EPA Code approved.
Cost: €4000-5000

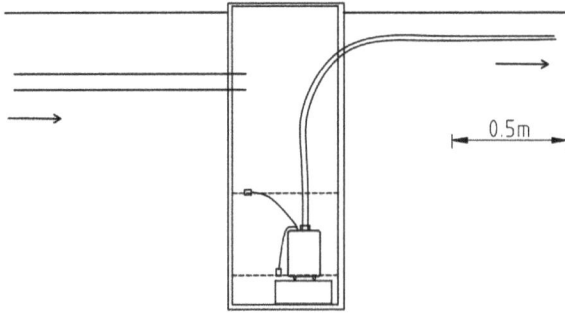

Pump Sump
c. 1m x 1m
Requires electricity.
Cost: €800

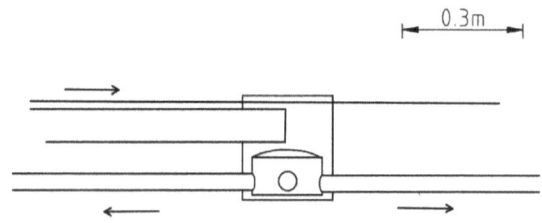

Ribbit Splitter Unit
Gravity distribution via 2; 4; 6; or 12 outlets.
Cost: €300 supply only

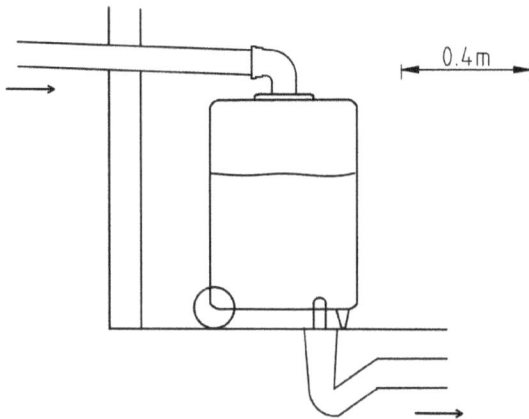

Settlement: Grey Water Woodchip Filter (grey water only)
c. 0.6m x 0.6m
Not EPA Code approved.
Cost: €150

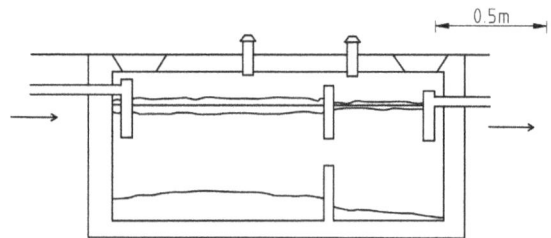

Settlement: Septic Tank
c. 3m x 2m; Request EPA approved size from supplier.
Cost: €1000-2000

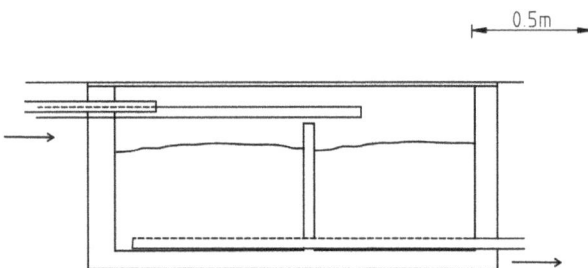

Settlement: Woodchip Filter Tank
c. 3m x 2m; takes both black water and grey water lines.
Not EPA Code approved (aka; Worm Tank in Australia or Brownfilter Tank by www.solviva.com USA).
Cost: €1-2000

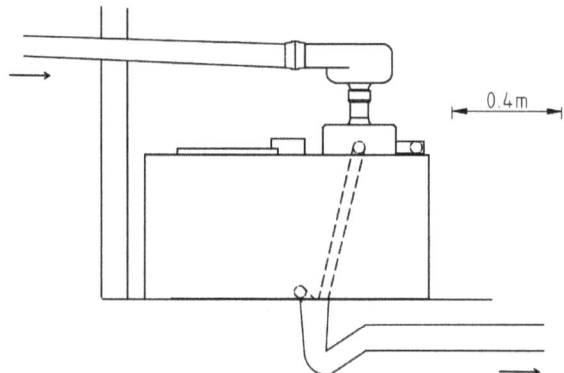

Settlement: Aquatron Faecal Separator (black water only)
c. 2m x 2m depending on collection chamber used. Requires site fall or electricity tp pump liquid to next stage.
Cost: €700 for separator – €5000 for chamber (optional)

Enjoy this book?

Permanent Publications is a small permaculture enterprise and ordering your books direct is like shopping locally.

Tell your friends! Your positive recommendations hugely help us reach a world in desperate need of positive and practical solutions.

We publish a range of books to empower and inspire changemakers the world over, from no dig organic growing, food forests and permaculture, to natural building, renewable technology and connecting with nature.

If you enjoyed *Septic Tank Options and Alternatives*, why not try these related titles.

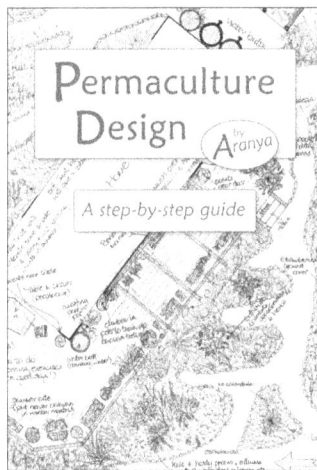

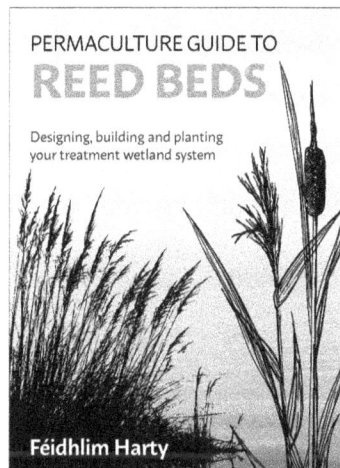

www.ingramcontent.com/pod-product-compliance
Lightning Source LLC
Chambersburg PA
CBHW050936210326
41518CB00024BB/2599